Modeling and Control of Infectious Diseases in the Host

Modeling and Control of Infectious Diseases in the Host

With Matlab and R

Esteban A. Hernandez-Vargas

Frankfurt am Main, Germany

Series Editor

Edgar Sánchez

ACADEMIC PRESS

An imprint of Elsevier

Academic Press is an imprint of Elsevier
125 London Wall, London EC2Y 5AS, United Kingdom
525 B Street, Suite 1650, San Diego, CA 92101, United States
50 Hampshire Street, 5th Floor, Cambridge, MA 02139, United States
The Boulevard, Langford Lane, Kidlington, Oxford OX5 1GB, United Kingdom

Notices

Knowledge and best practice in this field are constantly changing. As new research and experience broaden our understanding, changes in research methods, professional practices, or medical treatment may become necessary.

Practitioners and researchers must always rely on their own experience and knowledge in evaluating and using any information, methods, compounds, or experiments described herein. In using such information or methods they should be mindful of their own safety and the safety of others, including parties for whom they have a professional responsibility.

To the fullest extent of the law, neither the Publisher nor the authors, contributors, or editors, assume any liability for any injury and/or damage to persons or property as a matter of products liability, negligence or otherwise, or from any use or operation of any methods, products, instructions, or ideas contained in the material herein.

Library of Congress Cataloging-in-Publication Data
A catalog record for this book is available from the Library of Congress

British Library Cataloguing-in-Publication Data
A catalogue record for this book is available from the British Library

ISBN: 978-0-12-813052-0

For information on all Academic Press publications
visit our website at https://www.elsevier.com/books-and-journals

Working together
to grow libraries in
developing countries

www.elsevier.com • www.bookaid.org

Publisher: Mara Conner
Acquisition Editor: Chris Katsaropoulos
Editorial Project Manager: Peter Adamson
Production Project Manager: Paul Prasad Chandramohan
Designer: Matthew Limbert

Typeset by VTeX

To the entirely merciful, God
to the consciousness of my life, my father and mother
and to the origin and end of my deepest thoughts, my wife.

Esteban Abelardo

Contents

About the Author

Esteban A. Hernandez-Vargas is a research group leader and principal investigator at the Frankfurt Institute for Advanced Studies, Germany. In 2011, he obtained his PhD in Mathematics from the Hamilton Institute at the National University of Ireland. During three years, he hold a postdoctoral scientist position at the Helmholtz Centre for Infection Research (HZI) in Braunschweig, Germany. In July 2014, he founded the pioneering research group of Systems Medicine of Infectious Diseases at the HZI. Since March 2017, he and his research group moved to the Frankfurt Institute for Advanced Studies. Furthermore, he was a visiting scholar at Los Alamos National Laboratory (USA) and Universidad de Guadalajara (Mexico) and adjunct lecturer at the Otto von Guericke Universität Magdeburg, Germany. He has written more than 50 peer-reviewed publications. He is a member of the IEEE Control Systems Society, the Society of Mathematical Biology, the European Society of Virology, and the Mexican Research Council (CONACYT). Additionally, he is an Editorial Board member of Infectious Diseases of Frontiers in Medicine and Public Health Journal as well as a Research Topic Editor at Frontiers of Microbiology and Frontiers of Immunology. He is also involved in a number of IEEE and IFAC Conference Organizing Committees.

Preface

Modeling infectious diseases have undergone substantial development over the past two decades. Considerable efforts were initially invested to model the basic interactions of chronic viral infections with their respective target cells, for example, HIV infection. However, more recently, acute infectious diseases such as influenza and Ebola have largely increased the attention of the modeling community.

Once model parameters are fitted with experimental data, mathematical models have the potential to serve as virtual clinical trials to uncover the success and pitfalls of antiviral treatments as well as vaccines. Coupling within-host infection models, PK/PD dynamics, and control theoretical approaches are having a promising future to tailor treatments at a host level but also to implement public health policies to tackle infectious diseases pandemics and outbreaks.

This book reflects recent developments in infectious diseases modeling while providing a comprehensive introduction to multidisciplinary fields like virology, immune system, epidemiology, statistics, parameter fitting, and control systems. The first two parts of the book are recommended for teaching mathematical biology to undergraduates. Basic knowledge of programming and linear algebra are helpful though not essential as the book includes a self-contained introduction. The first part of the book (Chapters 1–3) brings together the basics necessary for modeling a broad of biological applications. The second part (Chapters 4–5) presents mathematical models for acute infections such as influenza and Ebola. Additionally, modeling HIV infection and its respective evolution during treatment is described in Chapter 6 and 7. For graduate students and scientists working in the control field of infectious diseases, the third part of the book (Chapters 8–10) is recommended, where more rigorous mathematical analysis is introduced. The book is supported by additional material including Matlab and R codes. Other advanced applications together with its codes can be found in https://github.com/systemsmedicine.

Esteban A. Hernandez-Vargas
Frankfurt am Main, Germany
August 2018

Acknowledgments

A central element to prepare this book is attributed to my students and the members of my research group of Systems Medicine of Infectious Diseases, in particular, Kinh Van Nguyen, Alessandro Boianelli, and Gustavo Hernandez Mejia. I feel truly fortunate to have had the opportunity to work with many people whose collaborations and discussions enriched the material presented in this book, including Richard Middleton (Newcastle University, Australia) and Patrizio Colaneri (Politecnico di Milano, Italy).

Furthermore, I acknowledge different funding institutions that supported my research positions at different stages such as the Alfons und Gertrud Kassel-Stiftung, the Bohringer Ingelheim Exploration Grant, the Helmholtz Initiative on Personalized Medicine Grant, the DAAD German Academic Exchange Service, the Mexican National Council of Science and Technology CONA-CYT, and the Science Foundation of Ireland. Finally, I thank the editor Edgar Sánchez and the Elsevier team composed by Chris Katsaropoulos and Katie Chan for guidance throughout the publishing process. I am also grateful to all the anonymous referees for carefully reviewing and selecting the final chapters of this book.

Esteban A. Hernandez-Vargas
Frankfurt am Main, Germany
December 2018

Part 1

Theoretical Biology Principles

Introduction

CONTENTS

1.1 MODELING AND CONTROL OF INFECTIOUS DISEASES

Throughout history, we have witnessed alarming high death tolls derived from infectious diseases around the globe [1]. One of the deadliest natural disasters in human history was caused by a viral infection, the 1918 flu pandemic, which killed approximately 50 million people. Infectious diseases are latent threats to humankind, killing annually 16 million people worldwide [2]. The magnitude of the threat represented by emerging virus diseases is immense; for example, HIV/AIDS, characterized in the early 1980s, has resulted in more than 30 million deaths from all socio-economic backgrounds. Furthermore, reemerging viruses like Ebola in 2014 and Zika in 2016 have baffled us with their threat to humans and health care systems around the globe. The consequences of epidemics can be devastating, infecting not only thousands of people but also animal populations and the food chain.

Antivirals and antibiotics are powerful weapons to fight against viral and bacterial infections, respectively. However, the misuse and overuse of antivirals and antibiotics have led to drug-resistant strains. In this respect, the World Health Organization (WHO) has reported that antimicrobial resistance (AMR) is a global health problem [3]. This has been clearly exposed by a growing list of bacteria that are becoming harder to treat due to antibiotics becoming less effective, for example, pneumococcus, staphylococcus aureus, mycobacterium tuberculosis, among others. The problem is also present in viral infections; for instance, HIV resistance has been reported with a range 7–40% in patients under antiretroviral therapies [4].

Modeling and Control of Infectious Diseases in the Host. https://doi.org/10.1016/B978-0-12-813052-0.00011-7

Several aspects that are critical for the development of effective therapeutic and prophylactic approaches to tackle infectious diseases remain largely fragmented. Optimal drug doses of antibiotics and antivirals have been performed integrating clinical observations and empirical knowledge to treat patients. However, this trial-and-error approach is becoming more challenging and infeasible by the steep increase in the amount of different pieces of information and the complexity of large datasets. Therefore, a systematic and tractable approach that integrates a variety of biological and medical research data is crucial to harness knowledge against infections.

Combining experimental and clinical data into mathematical models has played a key role to dissect underlying mechanisms that can lead to severe infections. In fact, the value of a model-based approach to drug development for improved efficiency and decision-making at preclinical development phases has been largely advocated by pharmaceutical companies [5].

The benefits of mathematical modeling are not only on generating and validating hypotheses from experimental data, but also simulating these models has led to testable experimental predictions [6]. Quantitative modeling is among the renowned methods used to study several aspects of viral infection diseases from virus–host interactions to complex immune responses systems and therapies [7–9]. However, incomplete and inadequate assessments in parameter estimation practices could hamper the parameter reliability, and consequently the insights that ultimately could arise from a mathematical model. To keep the diligent works in modeling biological systems from being mistrusted, potential sources of error must be acknowledged [10].

Joint forces between mathematical modeling and control theory can be instrumental to evaluate dose-response and predict the effect-time courses resulting from specific treatment. The design of optimal therapies is urgently needed to reduce the emergence of resistance-conferring mutations that can result in viral strains with reduced susceptibility to one or more of the drugs.

Through book chapters, existing modeling techniques, parameter estimation practices, and control theoretical approaches will be introduced to different infectious diseases such as HIV, influenza, and Ebola. In particular, the combination of optimal and impulsive control techniques are pivotal for the design of optimal antiviral therapies. The results that appear in next chapters include not only rigorous mathematical analyses but also simulation results to verify the performance of the corresponding schemes.

1.2 **BASICS OF IMMUNOLOGY**

Immunology is the branch of biology dedicated to the study of the immune responses derived in a host against diseases. Immunology is a very large research field that would require several courses and textbooks. Here, a very brief description of immunology and basic concepts are presented. The reader interested for deeper studies of immunology can consider the excellent reference popularly known between the immunologists as the "Janeway's" [11].

1.1 IMMUNE SYSTEM COMPONENTS

Roughly speaking, the immune system can be divided into two main parts, the innate and adaptive immune system. When an antigen, for instance a pathogen, is recognized by cells of the innate immune system, fast and non-specific mechanisms take place to clear the pathogen in the early hours post recognition. Activated by the innate system, adaptive immune responses are tailored to clear specifically the pathogen or infected cells. To avoid harming the host, the immune system has regulatory mechanisms to switch down the innate system and itself. A diagram illustrating these interactions is shown in Fig. 1.1. Note that antigens are molecules often external to the host although sometimes can be part of the host itself.

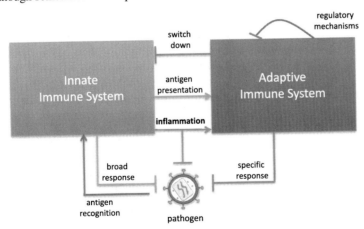

■ **FIGURE 1.1 Immune System Components.** The arrows represent activation, whereas connections ending with a vertical line represent inhibition.

Inflammation is a master regulatory process derived by the innate immune system to clear the initial cause of cell injury, to accelerate adaptive immune responses, and to initiate tissue repairment [12]. However, inflammation can be impaired and derived into chronic inflammation causing devastating events to the host health.

The innate immune system is composed of physiological barriers that prevent the invasion of foreign agents. Most infectious agents activate the innate immune system and induce an inflammatory response. The adaptive immune response is mediated by a complex network of specialized cells and chemical molecules that identify and respond to foreign invaders. The name of adaptive is due to the fact that can respond with great specificity to a very broad class of foreign substances, and can exhibit memory, which means that a subsequent rechallenge results in a conclusive immediate response.

The immune system is composed of different types of white blood cells (leukocytes), antibodies and active chemicals. Blood cellular elements such as the red blood cells that transport oxygen, the platelets that trigger blood clotting in damaged tissues, and the white blood cells are derived from the pluripotent hematopoietic stem cells in the bone marrow [11]. On the other hand, white cells are migratory agents that guard peripheral tissues, draining extracellular fluid that can free cells from tissues, transport them through the body as the lymph, and eventually into the blood.

A feature of many leukocytes is the presence of granules. **Granulocytes** are leukocytes characterized by the presence of differently staining granules in their cytoplasm. These granules are membrane-bound enzymes, which primarily act in the digestion of endocytosed particles. There are three main types of granulocytes such as neutrophils, basophils, and eosinophils. The most abundant type of granulocytes in most mammals are the neutrophils (40%–75%). Neutrophils can clear a large variety of microorganisms by phagocytosis, employing degradative enzymes and other antimicrobial substances stored in their cytoplasmic granules. Eosinophils and basophils are also in the primary defense against parasites; however, these can also contribute to allergic inflammatory reactions [11].

Agranulocytes are leukocytes characterized by the absence of granules in their cytoplasm, for example, lymphocytes and monocytes [11]. The three major types of lymphocyte are T cells, B cells, and natural killer (NK) cells. T cells (maturation in the Thymus) and B cells (maturation in Bone marrow) are the major cellular components of the adaptive immune response. Lymph nodes are principal sanctuaries of B and T lymphocytes, and other white blood cells. The specific roles played by various agranulocyte cells and their interactions are initiated and amplified by the induction of inflammatory responses of sensor cells. Neutrophils, dendritic cells, and macrophages are the main sensor cells to detect antigens and start innate immune responses.

Dendritic cells (DCs) are known as antigen-presenting cells (APCs), which are particularly important to activate T cells. Unlike others sensor cells, DCs do not kill directly infected cells but generate peptide antigens that can activate T cells. They have arms that look like dendrites of nerve cells (Fig. 1.2), hence their name. DCs express intracellular Toll-like receptors (TLR) that can recognize molecules derived from microbes. There are at least two broad classes of dendritic cells (DCs); the conventional dendritic cells (cDC) that seem to participate most directly in antigen presentation and activation of naive T cells. The second type are the plasmacytoid dendritic cells (pDCs), a distinct lineage that generates large amounts of interferons, particularly in response to viral infections, but do not seem to be as important for activating naive T cells. Once dendritic cells recognize antigens at the place of the infection, they migrate to the lymph nodes to activate T cells.

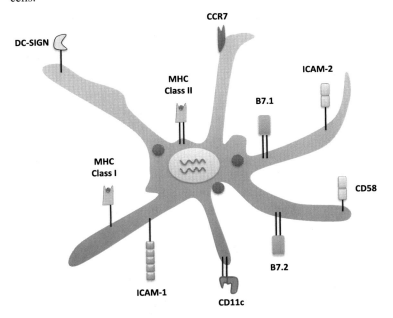

■ **FIGURE 1.2 Dendritic cell scheme.** DC-SIGN (Dendritic Cell-Specific Intercellular adhesion molecule-3-Grabbing Nonintegrin) also known as CD209 is a protein that serves as a receptor for several viruses. TLR-7 and TLR-9 are intracellular receptors for sensing viral infections; these TLRs can induce expression of the receptor CCR7. Mature DCs express MHC proteins for priming naive T cells and consequently can bind naive T cells using the intercellular adhesion molecules (ICAM-1 and ICAM-2) and the cell adhesion molecule CD58. The peripheral membrane proteins B7.1 and B7.2 of activated DCs can pair with either a CD28 or CD152 (CTLA-4) surface protein on T cells. CD11c protein induces cellular activation and triggers neutrophil burst.

Immune system cells detect peptides derived from foreign antigens. Such antigens peptide fragments are captured by **Major Histocompatibility**

Complex **(MHC)** molecules, which are displayed at the cell surface for recognition by the appropriate T cells. There are two main types of MHC molecules, MHC class I and class II. The key difference between these two classes lies not in their structure but in the source of the peptides that they can trap and carry to the cell surface. MHC class I molecules collect peptides derived from proteins synthesized in the cytosol; therefore they are able to display fragments of viral proteins on the cell surface. MHC class II molecules bind peptides derived from proteins in intracellular vesicles and thus display peptides derived from pathogens living in macrophage vesicles and B cells. Note that MHC class II receptors are only displayed by APCs [11].

Macrophages, large mononuclear phagocytic cells, are resident in almost all tissues and are the mature form of monocytes (Fig. 1.3). Macrophages circulate in the blood and continually migrate into tissues to differentiate. Macrophages are long-lived cells and perform different functions throughout the innate response and the subsequent adaptive immune response. Macrophages engulf and then digest cellular debris and pathogens (phagocytosis). Either as stationary or mobile cells, macrophages can stimulate lymphocytes and other immune cells to respond to pathogens [11].

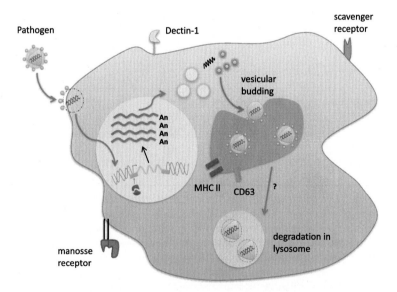

■ **FIGURE 1.3 Macrophage scheme.** Macrophages can express several cell-surface receptors that stimulate the phagocytosis and intracellular killing of microbes. Dectin-1 recognizes common components of fungal cell walls. Mannose receptors can recognize various ligands presented on bacteria, viruses, and fungi.

A crucial role of macrophages is to orchestrate the immune responses by inducing inflammation and secreting signaling proteins (cytokines and chemokines) that activate other immune system cells. **Cytokines** are secreted proteins by immune system cells that affects the behavior of nearby cells bearing appropriate receptors. **Chemokines** are secreted proteins that attract cells bearing chemokine receptors out of the blood stream and into the infected tissue.

T lymphocytes or T cells are a subset of lymphocytes defined by their development in the thymus. During T cell arrangement, a number of random rearrangements occur in the genome responsible for creating the T cell Receptors (TCRs) protein. This occurs in every immature T cell expressing a unique TCR surface protein, providing an enormous range of specificity. T cells that express TCRs that do not bind with sufficient strength to MHC molecules are eliminated (positive selection), as well as cells with strongly self-reactive receptors (negative selection). T cells that survive are those whose TCR proteins are capable of recognizing MHC molecules with bound peptide fragments but do not bind strongly to any peptide fragments occurring naturally in uninfected cells [11]. The lineage of the T cells is also determined in this selection process. T cells become either helper T cells (T_H or CD4+ T) expressing the surface molecule CD4 and TCR that bind with MHC class II or cytotoxic T cells (CTL or CD8+ T) expressing the surface molecule CD8 and TCR that bind to MHC class I; see Fig. 1.4.

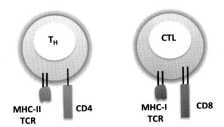

■ **FIGURE 1.4 Types of T cells.** On the left, it is shown a T cell expressing the surface molecule CD4, named helper T cells (T_H or CD4+ T). On the right, it is presented a T cell expressing the surface molecule CD8, named cytotoxic cells (CTL or CD8+ T).

Naive T cells are mature recirculating T cells that have not yet encountered their specific antigens. To participate in the adaptive immune response, a naive T cell must meet its specific antigen. Then the naive T cell becomes activated, which can proliferate and differentiate into cells that have acquired new activities that contribute to eliminate the antigen. Remarkably, CD4+ T cells do not directly participate in the clearance of the pathogen; instead, they regulate the development of the humoral (B-cell mediated) and cellular (CD8+ T cell mediated) immune responses.

CD8+ T cells carry the coreceptor CD8 and can recognize antigens that are synthesized in the cytoplasm of a cell. Naive CD8+ T cells are long-lived cells that remain dormant until interact with an APC displaying the antigen MHC class I complex. The costimulatory molecule B7 is also necessary to activate a CD8+ T cell into a cytotoxic-T cell (CTL), the quality of being toxic to cells. CTL recognizes and binds virus-infected cell inducing apoptosis in the target cell. A naive T cell can rapidly undergo up to 15 divisions to produce more than 10^4 daughter cells within one week [13]. CD8+ T clonal expansion and survival highly depend on CD4+ T cells, although the underlying mechanisms are still debated [13].

Natural Killers (NKs) are another type of cytotoxic cells but from the innate immune system that can recognize and kill cells infected by some pathogens. The name is attributed due to NKs do not require activation to kill cells that are missing self-markers of MHC I. Thus NKs are particularly important to detect and destroy harmful cells that are missing MHC I markers. NK cells are essential for keeping many viral infections in check before CD8+ T cells become functional [14].

CD4+ T cells have a more flexible repertoire of effector activities than CD8+ T cells. T_H1 cells are able to activate infected macrophages and CD8+ T cells. T_H2 cells provide help to B cells for antibody production. T_H17 cells promote resistance to extracellular bacteria and fungi. T_{FH} cells promote the formation and maintenance of germinal centers mediating the selection and survival of B cells. T_{reg} cells suppress the immune system, maintain tolerance to self-antigens, and prevent autoimmune disease [11]. As a note, tolerance can be roughly defined as the nonreactivity respect to an antigen that could normally be expected to promote an immunological response. Naive T_H can differentiate into different types as previously mentioned, and this process is remarkably plastic, where cytokines are master regulators [14]. Cytokines have a central role in the development, differentiation, and regulation of immune cells [14]; see Fig. 1.5. A detailed mapping of the immune system is crucial to combat diseases; however, the immune system is complex and remains largely unknown. The reduced cell-cytokine network presented in Fig. 1.5 is thus just the tip of the iceberg.

Lymphocytes are located in different parts in the human body. They can circulate through the primary lymphoid organs (thymus and bone marrow), the secondary lymphoid organs (spleen, lymph nodes (LN)), tonsils and Peyers patches (PP) as well as nonlymphoid organs, such as blood, lung, and liver; see Table 1.1. Lymphocytes cell counts in the blood are used to evaluate the immune status because it is an accessible organ system; however, blood lymphocytes represent only about 2% of the total numbers of lymphocytes

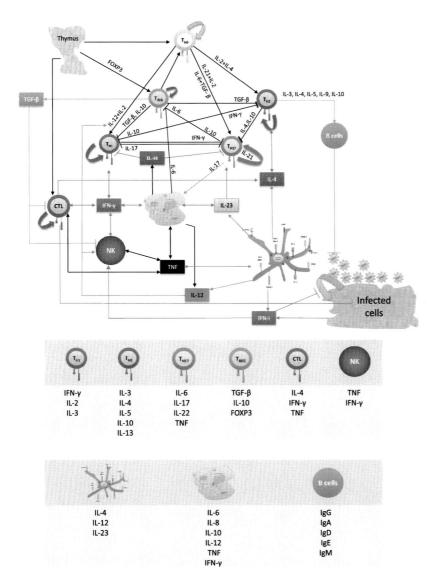

T_{H1}	T_{H2}	T_{H17}	T_{REG}	CTL	NK
IFN-γ	IL-3	IL-6	TGF-β	IL-4	TNF
IL-2	IL-4	IL-17	IL-10	IFN-γ	IFN-γ
IL-3	IL-5	IL-22	FOXP3	TNF	
	IL-10	TNF			
	IL-13				

		B cells
IL-4	IL-6	IgG
IL-12	IL-8	IgA
IL-23	IL-10	IgD
	IL-12	IgE
	TNF	IgM
	IFN-γ	

in the body. The number of lymphocytes in the blood depend on race and is influenced by various factors like race and age [15].

The humoral immune response is the transformation of B cells into plasma cells that can produce and secrete antibodies. **B cells** are primarily involved in the production of antibodies, proteins that bind with extreme specificity to a variety of extracellular antigens. B cells (with costimulation) transform into plasma cells with secrete antibodies. The costimulation of the B cell can

■ **FIGURE 1.5 Reduced cell-cytokines Network.** The arrows represent activation, whereas connections ending with a vertical line represent inhibition. The fate decision of CD4+ T cells to differentiate in any of the subsets is represented as a triangular form in the upper part of the network. The lower table shows the different immune system cells and their respective produced components that can modulate other parts of the immune system.

Table 1.1 Lymphocytes distribution	
Organ	**Lymphocytes (10^9)**
Blood	10
Lung	30
Liver	10
Spleen	70
Lymph nodes	190
Gut	50
Bone marrow	50
Thymus	50
Source: [15].	

come from another APC, like a dendritic cell. This entire process is aided by T_H2 cells and T_{FH}, which provide costimulation. Antibody molecules are known as immunoglobulins (Ig), and the antigen receptor of B lymphocytes as membrane immunoglobulin. Antibodies can protect against infections by *i)* preventing the pathogen to interact with the host cells; *ii)* neutralizing complete pathogen particles by binding and inactivating them; and *iii)* coating antigens to be recognizable as foreign by phagocytes [14].

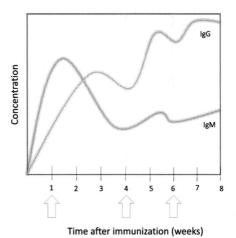

■ **FIGURE 1.6 Antibody dynamics after vaccination.** Affinity maturation and high production of antibodies occur 2–3 weeks after the first immunization. Vertical arrows represent the time when boosters were provided.

There are five classes of antibodies that B cells can produce: IgM, IgG, IgA, IgD, and IgE. Each antibody has different chemical structures in their invariant region (the portion of the molecule that does not affect the antibodies antigen specificity). Antibodies can appear either after infections (active immunization) or after antigenic material stimulation by vaccines (passive immunization). IgM is a temporary antibody that disappears within 2–3 weeks while is replaced by IgG, which provides lasting immunity [14]. Antibody response can be characterized by an initial rapid production of IgM followed by a slightly delayed IgG response (Fig. 1.6). After repeated immunizations (boosters), much larger amounts of IgG antibody are present.

1.3 BASICS OF VIROLOGY

Virology is the branch of biology dedicated to study viruses, its structures mechanisms of replication inside of host's cells, and methods used to analyze viruses. An excellent reference for further concepts and techniques of virology can be found in [16]. **Viruses** are small infectious nonliving agents that can enter host cells like a Trojan horse and hijack their molecular machinery to replicate inside of cells. It is estimated that there are around 10^{31} viruses on earth [2]; many of viruses are harmless, while others represent a worldwide threat to health. The virus particles, also known as *virions*, contain the genetic material necessary to replicate new copies for progeny virions; a generalized replication cycle is broadly explained in Box 1.2.

1.2 VIRUS REPLICATION

It is a common mistake in scientific fields outside of virology to say that a virus grows or a virus is reproduced; the right term is a virus replicates. Virus replication is the mechanistic process of a virus to make copies of itself inside of a host cell. Although different viruses can employ different pathways or mechanisms, a general viral replication cycle described in [16] is summarized in the following steps:

Step 1: Attachment of a virus to a cell. Viruses use surface proteins (receptors) to specific molecules located on the surface of the host cell. Note that a virus initially binds to some cell molecules, and if remains, then the virus can continue binding to more receptors, which is considered a strong binding and irreversible.

Step 2: Entry of a virus into the cell. After the attachment, the virus requires to cross the plasma membrane to release the viral genome (also known as uncoating).

Step 3–4: Transcription and translation. Transcription is the process by which DNA is copied into a messenger RNA (mRNA) molecule. Consequently, the mRNA is transported out of the nucleus into the cytoplasm. The mRNA is then translated in a sequence of amino acids during protein synthesis, in this case into virus proteins. This transcription–translation process is known as the central dogma of molecular biology proposed by Francis Crick in 1958 [17]. The central dogma was modified and generally accepted as follows:

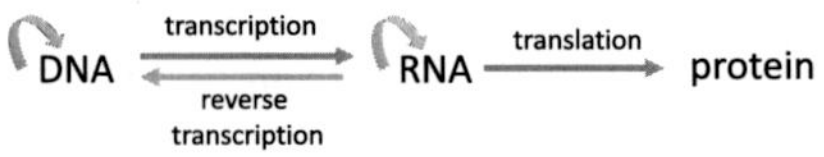

$$\text{DNA} \underset{\substack{\text{reverse}\\ \text{transcription}}}{\overset{\text{transcription}}{\rightleftharpoons}} \text{RNA} \xrightarrow{\text{translation}} \text{protein}$$

Step 5: Genome Replication. The genome is also replicated in this manner, and then viral transcription is triggered providing more copies of the genome.

Step 6: Virion Assembly. New synthesized genome and proteins are assembled to form new virions.

Step 7: Release. Mature virions are released through the cell membrane remaining inert till they encounter a new host cell.

Viruses are different not only in their structure but also in their genome and way of replication, which gives the basis for classifying viruses in seven groups (Fig. 1.7). Depending on the type of infection, viruses can be transmitted to humans from infected humans, animals, or via vectors (e.g., insects, mites, ticks). Once new cycles of replication take place, it is considered as successfully transmitted. Note that although one single viral particle could initiate an infection, in practice a minimum number of virions are required to initiate an infection.

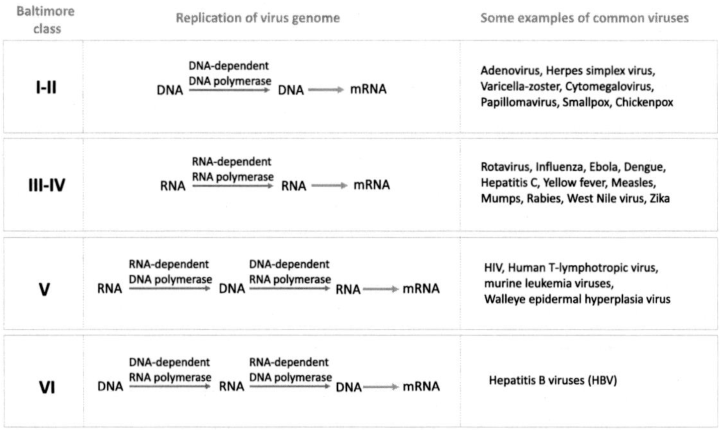

Baltimore class	Replication of virus genome	Some examples of common viruses
I-II	DNA $\xrightarrow[\text{DNA polymerase}]{\text{DNA-dependent}}$ DNA $\longrightarrow$ mRNA	Adenovirus, Herpes simplex virus, Varicella-zoster, Cytomegalovirus, Papillomavirus, Smallpox, Chickenpox
III-IV	RNA $\xrightarrow[\text{RNA polymerase}]{\text{RNA-dependent}}$ RNA $\longrightarrow$ mRNA	Rotavirus, Influenza, Ebola, Dengue, Hepatitis C, Yellow fever, Measles, Mumps, Rabies, West Nile virus, Zika
V	RNA $\xrightarrow[\text{DNA polymerase}]{\text{RNA-dependent}}$ DNA $\xrightarrow[\text{RNA polymerase}]{\text{DNA-dependent}}$ RNA $\longrightarrow$ mRNA	HIV, Human T-lymphotropic virus, murine leukemia viruses, Walleye epidermal hyperplasia virus
VI	DNA $\xrightarrow[\text{RNA polymerase}]{\text{DNA-dependent}}$ RNA $\xrightarrow[\text{DNA polymerase}]{\text{RNA-dependent}}$ DNA $\longrightarrow$ mRNA	Hepatitis B viruses (HBV)

■ **FIGURE 1.7 Virus Classification.** The Baltimore classification places the viruses in seven groups, this depends of their nucleic acid (DNA or RNA), strandedness (single-stranded or double-stranded), and method of replication; further details in [16].

Viruses can form visible zones in layers of host cells named as *plaques*, which provide a way to quantify the concentration (also called viral titer) of infective virus in plaque-forming units (pfu) [16]. Viruses need to infect other cells to continue the replication process; however, many virions do not have the ability to be *infective*. There are several assays to quantify viruses. The standard method is to count the numbers of plaques in samples with different virus dilutions (pfu/ml). Another assay is the $TCID_{50}$ (50% Tissue culture Infective Dose), which quantifies the amount of virus required to kill 50% of infected hosts. Note that 1 $TCID_{50}$ can be approximated as 0.7 pfu.

1.4 **VIRAL MUTATION AND DRUG RESISTANCE**

Viral replication is an error-prone process promoting genetic modifications. These errors, known as *mutations*, not only affect the genotype (set of genes) but also modify observable characteristics of the virus (phenotype). Genotypic and phenotypic resistance assays are used to assess viral strains and inform selection of treatment strategies. On one hand, the genotype is the genetic makeup of a cell or an organism. The genotypic assays detect drug resistance mutations present in relevant genes. On the other hand, a phenotype is any observable characteristic or trait of an organism, such as its morphology, development, biochemical or physiological properties, and behavior. The phenotypic assays measure the ability of a virus to grow in different concentrations of antiretroviral drugs [18].

Mutation rate is a way to measure the number of substitutions or changes in the sequence of a gene that occur over time or per cell infection cycle. Mutation rates vary among different viruses from 10^{-8} to 10^{-3} substitutions per nucleotide per cell infection (s/n/c). RNA viruses have a mutation rate higher than DNA viruses, that is, 10^{-6}–10^{-3} for RNA viruses whereas 10^{-8}–10^{-6} for DNA viruses [19]. Viral mutation rates are dynamic in response to specific selective pressures, that is, immune response or antiviral treatment. The problem resides on the fact that these new viral variants have the potential to evade the immune system responses or generate resistance to antiviral treatments [19]. A summary of viral mutation rates of some common viruses is presented in Table 1.2.

For example, drug resistance is a critical problem in HIV infection, resulting in the reduction in effectiveness of a drug in curing the disease. Attributed to the reverse transcription process, HIV differs from many viruses because it has very high genetic variability. This diversity is a result of its fast replication cycle, with the generation of about 10^{10} virions every day, coupled with a high mutation rate of approximately 3×10^{-5} per nucleotide base per

Table 1.2 Summary of mutation rates

Virus	Mutation rate (s/n/c)[a]	Mutation rate (s/s/y)[b]
Hepatitis C virus	3.8×10^{-5}	0.79×10^{-3}
Influenza A virus	2.5×10^{-5}	1.80×10^{-3}
Measles virus	3.5×10^{-5}	0.40×10^{-3}
HIV-1 (free virions)	6.3×10^{-5}	2.50×10^{-3}
HIV-1 (cellular DNA)	4.4×10^{-3}	n.a.
Human cytomegalovirus	2.0×10^{-7}	n.a.
Human rhinovirus	6.9×10^{-5}	0.58×10^{-3}
Dengue 4	–	0.77×10^{-3}
Ebola	–	$1.30 \times 10^{-3(c)}$
Zika	–	$0.61 \times 10^{-3(d)}$
Chikungunya	–	$0.43 \times 10^{-3(e)}$

(∗) *substitutions per nucleotide per cell infection (s/n/c).*
(∗) *substitutions per site per year (s/s/y).*
[a] *Geometric mean calculated in [19].*
[b] *Source: [20].*
[c] *Source: [21].*
[d] *Source: [22].*
[e] *Source: [23].*

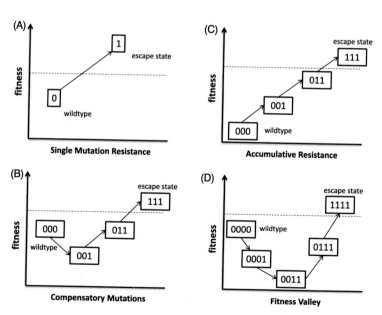

■ **FIGURE 1.8 Resistance pathways.** Depending on the virus class, resistance could occur by a single mutation (A), compensatory mutations (B), accumulative mutations (C), or by a fitness valley shape mutation (D). The genome is represented by the binary string form by 0 and 1 inside of the squares. Further details in [25].

cycle of replication [24]. This complex scenario leads to the generation of many variants of HIV in the course of one day.

Fitness can be defined as the capacity of a virus to replicate an infectious progeny in a given environment [26]. Even small increases in mutation rates of RNA viruses can result in serious fitness effects [27]. The frequencies of the genotypes will change over generations, and only those variants with higher fitness will dominate the population. Drug resistance can be attributed to single or multiple mutations, enabling the emergence of variants that are fit enough to sustain the population. Fig. 1.8A illustrates a resistance pathway with a single point mutation (e.g., single drug treatment in HIV infection). Drug resistance can also emerge through accumulation of resistance-associated mutations or after multiple steps of fitness; see Fig. 1.8B–D.

Chapter

2

Mathematical Modeling Principles

CONTENTS

2.1 MATHEMATICAL MODELING

Mathematical modeling is an abstract representation of a system based on mathematical terms in order to study the effects of different components and consequently to make predictions. Mathematical modeling approaches have been playing a central role in describing many different applications in engineering and natural and social sciences [28]. In biological and medical research, benefits from mathematical modeling have not been only on generating and validating hypotheses from experimental data, but also simulating these models has led to testable experimental predictions [6]. Nowadays, modeling is among the renowned methods used to study several aspects of viral infection diseases from virus–host interactions to complex immune response systems and therapies [7,9,29,30].

Mechanistic models aim to mimic biological mechanisms that embody some essential and exciting aspects of a particular disease. These can serve either as pedagogical tools to understand and predict disease progression or act as the objects of further experiments. Model selection is a central part of modeling that can be performed by fitting the models to experimental data and comparing the models' goodness-of-fit criterion [31–33]. Alternatively, modeling can be used as "thought experiments" without considering parameter fitting to experimental data. For instance, on the basis of various simple mathematical models, the hypothesis that the immune activation determines the decline of memory CD4+ T cells in HIV infection was rejected [34].

Mathematical models at heart are incomplete, and the same process can be modeled differently, if not competitively. Inherently, a model should not be accepted or proven as a correct one, but should only be seen as the least

Modeling and Control of Infectious Diseases in the Host. https://doi.org/10.1016/B978-0-12-813052-0.00012-9

19

invalid model among the alternatives. In the author's opinion, the most appropriate way to show the potential of mathematical modeling is by the following quote:

> *"All models are wrong, but specific assumptions can be useful for predicting experiments and explaining biological problems"*

Prof. Pedro Mendes

Assumptions, something accepted as true, are a critical component in mathematical modeling in order to reduce the complexity of a model. However, assumptions are generally valid under very specific and limited circumstances. If a model is inadequate or fails to represent the reality, then assumptions need to be reexamined. Cycles intertwining modeling and experiments are therefore needed to fine tune mathematical models as presented in Box 2.3. This chapter is dedicated to the mathematical formulation (step 1 in Box 2.3); parameter estimation and model analysis will be provided in the next chapters.

2.3 MODELING CYCLE

Based on the observed process, the formulation of mathematical models is developed (**step 1**). Using experimental data model parameters are then estimated (**step 2**). Note that multiple models can provide the same fit with observed experimental data. Thus, a model selection criterion is necessary. Parameter uncertainty is evaluated providing parameter confidence intervals and a sensitivity analysis to show the relation between parameters and the respective influence on the model outcome (**step 3**). Once parameter distributions are inferred, we can test the predictive power of the mathematical model leading to new knowledge of the observed process and guiding the design of new experiments (**step 4**).

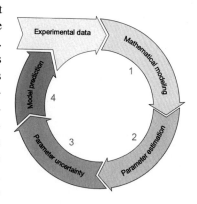

There are different classifications of mathematical models based on their structure (linear vs. nonlinear), time-dependent changes (static vs. dynamic), object consideration (continuous vs. discrete), relationship between variables in the data (mechanistic vs. black-box), scale (microscopic vs. macroscopic), precision (qualitative vs. quantitative), and randomness (deterministic vs. stochastic). This last classification is one of the most known ways to differentiate models. Deterministic models are fully determined by parameters in the model and by sets of previous states, giving always the same results for a given set of parameters and initial conditions. Conversely, stochastic models are not described by unique variable states but rather by probability distributions.

A modeler needs to take several decisions to some point arbitrary when a model is developed, such as model complexity and methodology. The process to be modeled can help to determine the methodology to be employed. For instance, deterministic models need several assumptions on the system such as high number of particles interacting in the process and well-mixed, free of noise conditions. On the other hand, stochastic models do not require any number of particles; these can work on dilute well-mixed conditions and consider noise signals. Although stochastic models can provide a more realistic approach for many applications, these are computationally demanding. For the case of infectious diseases, **population modeling** is the most common approach used to predict the population dynamics of pathogens within-hosts and between-hosts. This approach can help to understand not only the temporal isolated of one population but also how different populations can interact and affect each other. Understanding dynamic changes of populations is a central characteristic across infections, therefore the importance of differential equations. Notations, basic concepts of linear algebra, differential equations, and stability concepts are introduced for further sections presenting host infection models. The textbook by Gershenfeld [28] is recommended to the readers interested for further details of modeling topics and methodologies.

2.2 MATHEMATICAL PRELIMINARIES

Notation. Throughout, $\mathbb{R}$ denotes the field of real numbers, $\mathbb{R}^n$ stands for the vector space of all n-tuples of real numbers, and $\mathbb{R}^{n \times n}$ is the space of $n \times n$ matrices with real entries. By $\mathbb{N}$ we denote the set of natural numbers and by $\mathbb{C}$ the set of complex numbers. For x in $\mathbb{R}^n$, x_i denotes the ith component of x, and the notation $x \succeq 0$ means that $x_i \geq 0$ for $1 \leq i \leq n$. By $\mathbb{R}^n_+ = \{x \in \mathbb{R}^n : x \succeq 0\}$ we denote the nonnegative orthant in $\mathbb{R}^n$. For differentiation of an independent variable x, the Leibniz notation $\frac{dx}{dt}$ will be represented through the book more frequently for mathematical models, whereas the Newton notation $\dot{x}$ will be more often used for control engineering sections.

Next, the most important matrix operators are presented. The material introduced in this section is based on the following books dealing with matrix theory and linear algebra: Lancaster [35] and Poznyak [36].

Definition 2.1. A matrix is a rectangular array

$$\begin{bmatrix} a_{11} & a_{12} & \cdots & a_{1n} \\ a_{21} & a_{22} & \cdots & a_{2n} \\ \vdots & \vdots & \ddots & \vdots \\ a_{m1} & a_{m2} & \cdots & a_{mn} \end{bmatrix},$$

of numbers $(a_{i,j})$ for which operations such as addition and multiplication are defined.

The size of a matrix is defined by the number of rows and columns. A matrix with m rows and n columns is called an $m \times n$ matrix. A raw vector in $\mathbb{R}^m$ is a $1 \times m$ matrix

$$\begin{bmatrix} a_1 & a_2 & a_3 & \cdots & a_n \end{bmatrix}.$$

A matrix with equal number of rows and columns is commonly known as a squared matrix. Several operators are only defined for square matrices. A particular case of a squared matrix is the diagonal matrix for which all elements outside the main diagonal are equal to zero:

$$\begin{bmatrix} a_{11} & 0 & \cdots & 0 \\ 0 & a_{22} & \cdots & 0 \\ \vdots & \vdots & \ddots & \vdots \\ 0 & 0 & \cdots & a_{nn} \end{bmatrix} = diag[a_{11}, a_{22}, \ldots, a_{nn}].$$

Definition 2.2. The sum of two matrices A and B, both with dimension m by n, is calculated as $(A + B) = (A + B)_{i,j} = A_{i,j} + B_{i,j}$ for $1 \leq i \leq m$ and $1 \leq j \leq n$. Note that matrices with different dimensions cannot be calculated.

Definition 2.3. The product of a number c and a matrix A is computed by multiplying every entry of A by c, that is, $(cA)_{i,j} = cA_{i,j}$.

Definition 2.4. The multiplication of two matrices $A \in \mathbb{R}^{m \times n}$ and $B \in \mathbb{R}^{n \times p}$ can be defined if and only if the number of columns of the left matrix (A) is the same as the number of rows of the right matrix (B):

$$[AB]_{i,j} = \sum_{r=1}^{n} A_{i,r} B_{r,j}.$$

Note that the multiplication of matrices is generally not commutative, that is, $AB \neq BA$.

Definition 2.5. The transpose of a matrix $A \in \mathbb{R}^{m \times n}$ is formed by turning rows into columns and vice versa, that is, $A' \in \mathbb{R}^{n \times m}$.

Definition 2.6. A matrix $A \in \mathbb{R}^{n \times n}$ is normal if $AA' = A'A$.

Definition 2.7. A matrix $A \in \mathbb{R}^{n \times n}$ is symmetric if $A = A'$.

Definition 2.8. A matrix $A \in \mathbb{R}^{n \times n}$ is orthogonal if $AA' = A'A = I \in \mathbb{R}^{n \times n}$.

Definition 2.9. The Hermitian transpose of $A \in \mathbb{C}^{m \times n}$ denoted by A^* is the result by taking the transpose and then taking the complex conjugate of each entry, that is, negating the imaginary parts but not their real parts.

Definition 2.10. The trace of a matrix $A \in \mathbb{C}^{n \times n}$, written by $tr(A)$, is defined as the sum of all elements on the main diagonal of the matrix A.

Considering the set of vectors $x_1, x_2, \ldots, x_k \in \mathbb{C}^n$ and constants $\alpha_1, \alpha_2, \ldots, \alpha_k \in \mathbb{C}^n$, the following notions are defined.

Definition 2.11. The inner product of two vectors $x \in \mathbb{C}^{n \times 1}$ and $y \in \mathbb{C}^{1 \times n}$ is given by

$$(x, y) := x'y = \sum_{i=1}^{n} x_i y_i.$$

Definition 2.12. Linear combinations of $x_1, x_2, \ldots, x_k$ can be written in the form

$$\alpha_1 x_1 + \alpha_2 x_2 + \cdots + \alpha_k x_k.$$

Definition 2.13. The set of all linear combinations of $x_1, x_2, \ldots, x_k$ is called the subspace or the span of $x_1, x_2, \ldots, x_k$, denoted by

$$span\{x_1, x_2, \ldots, x_k\} := \{x = \alpha_1 x_1 + \alpha_2 x_2 + \cdots + \alpha_k x_k : \alpha_i \in i = 1, \ldots, k\}.$$

Definition 2.14. The vectors $x_1, x_2, \ldots, x_k$ are linearly dependent if there exist a parameter set of $\alpha_1, \alpha_2, \ldots, \alpha_k \in \mathbb{C}^n$, not all zero, such that

$$\alpha_1 x_1 + \alpha_2 x_2 + \cdots + \alpha_k x_k = 0.$$

Otherwise, they are said to be linearly independent.

Definition 2.15. The rank of a matrix $A \in \mathbb{C}^{n \times n}$, denoted by $rank(A)$, is the number of linearly independent rows or columns of a full matrix.

Definition 2.16. The image or range (Im) of a linear transformation $A : \mathbb{C}^n \rightarrow \mathbb{C}^m$ is

$$Im \ A := \{y \in \mathbb{C}^m : y = Ax, x \in \mathbb{C}^n\}.$$

Definition 2.17. A matrix $A \in \mathbb{C}^{m \times n}$ can be considered as a linear transformation from $\mathbb{C}^n$ to $\mathbb{C}^m$. In this fashion, the kernel (Ker), also known as the null space of the linear transformation $A : \mathbb{C}^n \rightarrow \mathbb{C}^m$, is defined by

$$Ker \ A := \{x \in \mathbb{C}^n : Ax = 0\}.$$

Definition 2.18. For any $A \in \mathbb{R}^{m \times n}$ and $B \in \mathbb{R}^{n \times p}$, the *Sylvester rule* states as follows:

$$
\begin{aligned}
rank(A) + rank(B) - n \quad &\leq \quad rank(AB) &\quad (2.1)\\
&\leq \quad min \ \{rank(A), rank(B)\}. &\quad (2.2)
\end{aligned}
$$

Furthermore, for square matrices $A \in \mathbb{R}^{n \times n}$ and $B \in \mathbb{R}^{n \times n}$, the following inequalities hold:

$$
\begin{aligned}
rank(A + B) &\leq rank(A) + rank(B),\\
rank(AB) &\leq min\{rank(A), rank(B)\},\\
dim \ Ker \ A + dim \ Im \ A &= n,\\
rank \ A &\leq dim \ Im \ A.
\end{aligned}
$$

Definition 2.19. The inverse (A^{-1}) of a squared matrix $A \in \mathbb{R}^{n \times n}$ has the property of $AA^{-1} = I \in \mathbb{R}^{n \times n}$. A square matrix that is not invertible is called singular or degenerate.

Remark 2.1. For a square $n \times n$ matrix, $rank(A) = n$ if and only if it is nonsingular. A square matrix is full rank if each of the rows of the matrix are linearly independent, that is, nonsingular. Nonsingular matrices have an inverse and a determinant different from zero.

Definition 2.20. The determinant is an operation computed only in a square matrix. In this book, the determinant of a matrix A is denoted as $det(A)$. For the case of a 2×2 matrix, the determinant can be computed as follows:

$$det(A) = \begin{vmatrix} a & b \\ c & d \end{vmatrix} = ad - bc.$$

2.4 EIGENVALUES AND EIGENVECTORS

In linear algebra, eigenvalues and eigenvectors are central features with a wide range of applications in mathematics, engineering, and computational biology. Roughly speaking, if a matrix A acts by stretching the vector x without changing its direction, then x is an eigenvector of A.

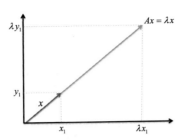

Definition 2.21. An eigenvector x of a square matrix $A \in \mathbb{C}^{n \times n}$ is a nonzero vector that multiplied by the matrix A remains parallel to the original vector x. A vector x is a right eigenvector of the matrix A if it satisfies $Ax = \lambda x$.

Definition 2.22. The eigenvalue λ is the corresponding factor by which the eigenvector is scaled when multiplied by the matrix A.

Remark 2.2. Distinct eigenvalues form linearly independent eigenvectors.

Definition 2.23. For any matrix $A \in \mathbb{C}^{n \times n}$, any eigenvalue λ satisfies the characteristic equation

$$p_A(\lambda) := det(\lambda I_{n \times n} - A) = 0.$$

The set of all roots of $p_A(\lambda)$ is called the spectrum and contains all right eigenvalues of the matrix A.

Definition 2.24. The set of all roots of $p_A(\lambda)$ is called the spectrum of A and is denoted by

$$\sigma(A) := \{\lambda_1, \lambda_2, \dots, \lambda_n\},$$

where λ_j satisfies $p_A(\lambda_j) = 0$.

Definition 2.25. The maximum modulus of the eigenvalues is called the spectral radius of A, denoted by

$$\rho^s(A) := \max_{1 \le j \le n} |\lambda_j|.$$

Remark 2.3. If a matrix A is symmetric, then all its eigenvalues λ_j are real.

Definition 2.26. Given a linear mapping $T : \mathbb{R}^n \to \mathbb{R}^m$, an invariant subspace W has the property that all vectors $v \in W$ are transformed by T into vectors also contained in W.

Remark 2.4. For a diagonal matrix $A = diag[\lambda_1, \lambda_2, \ldots, \lambda_j]$, the eigenvalues λ_j are the elements of its diagonal. Additionally, the $det A = \prod_{i=1}^{n} \lambda_j$.

Definition 2.27. For a matrix $A \in \mathbb{C}^{n \times n}$, the algebraic multiplicity of an eigenvalue λ_i is the number of times that λ_i appears repeated as a root of $p_A(\lambda)$.

Definition 2.28. Denoting the dimension of a subspace as dim, the geometric multiplicity of an eigenvalue λ_i is the number of linearly independent eigenvectors corresponding to λ_i and can be computed by

$$dim \ Ker \ (\lambda_i I_{n \times n} - A).$$

The geometric multiplicity of an eigenvalue does not exceed its algebraic multiplicity.

Definition 2.29. A matrix is *Hurwitz* if all its eigenvalues have negative real parts.

2.5 BRIEF MATRIX COOKBOOK

Given $A \in \mathbb{R}^{n \times n}$ and $B \in \mathbb{R}^{n \times n}$, the following matrix properties hold:

$$(A + B) + C = A + (B + C),$$

$$A^p = \underbrace{AA \ldots A}_{p \text{ times}},$$

$$(A + B)C = AC + BC,$$

$$(AB)C = A(BC),$$

$$A^p A^q = A^{p+q},$$

$$(AB)' = B'A',$$

$$(A^p)^q = A^{pq},$$

$$(ABC\ldots)' = \ldots C'B'A',$$

$$det(A^{-1}) = \frac{1}{det(A)},$$

$$det(AB) = det(A)det(B),$$

$$(AB)^{-1} = B^{-1}A^{-1},$$

$$det(A) = det(A'),$$

$$(A')^{-1} = (A^{-1})'.$$

$$det(AB) = det(A)det(B);$$

A more complete list of matrix identities and linear algebra can be found in [36].

2.3 DYNAMICAL SYSTEMS

Ordinary differential equations (ODEs) are equations containing functions of one independent variable and its respective derivatives. In this regard, ODEs play a central role to represent in mathematical term a wide range of complex problems in biology, engineering, physics, chemistry, and economics. In a general form, ODEs can be represented as follows:

$$\frac{dx}{dt} = f(x, t), \tag{2.3}$$

where $\frac{dx}{dt}$ represents the changes of the variable x with respect to the time by the variable t. The function $f : \mathbb{R}^n \to \mathbb{R}^n$ depends on x and t. The term "ordinary" is used in contrast with the term partial differential equation, which may be with respect to more than one independent variable and its partial derivatives. Linear ODEs are generally represented in a matrix form:

$$\frac{dx}{dt} = Ax, \tag{2.4}$$

where A is an $n \times n$ square matrix of the constant coefficients $a_{ij} \in \mathbb{R}^{m \times s}$.

Definition 2.30. A function F is linear if for any two vectors x and y and any scalar α, the *additivity property*

$$F(x + y) = F(x) + F(y)$$

and the *homogeneity property*

$$F(\alpha x) = \alpha F(x)$$

are satisfied. Nonlinear functions are all those that are not linear.

The key property of dynamical systems based on ODEs is the equilibrium point, represented by $x = x^*$. The meaning behind **equilibrium points** is that whenever the state of the system starts, it will remain at x^* for all future time. Note that if there are no other equilibriums points, then the equilibrium point is considered isolated; otherwise, there are several equilibrium points, and the system will converge to the closest stable equilibrium point.

In fact, a system in its steady state has numerous properties that are unchanging in time. The equilibrium points are the real roots of a system ($f(x, t) = 0$) implying that the partial derivative with respect to time is zero, that is, $\frac{\partial x}{\partial t} = 0$. In many systems, steady state is not reached until some time known as transient has elapsed after the system is initiated. Stability concepts presented in this section can be found in extended form in [37], which is a classic book for the study of nonlinear systems.

2.6 STABILITY

Roughly speaking, an equilibrium point is considered stable if all solutions starting at nearby points stay nearby. For the system $\dot{x} = f(x, t)$, stability for the equilibrium point $x = 0$ is defined as

- stable if, for each $\epsilon > 0$, there is $\delta = \delta(\epsilon)$ such that

$$\|x(0)\| < \delta \rightarrow \|x(t)\| < \epsilon \quad \forall t \geq 0,$$

- unstable if it is not stable,
- asymptotically stable if it is stable and δ can be chosen such that

$$\|x(0)\| < \delta \rightarrow \lim_{t \to \infty} x(t) = 0.$$

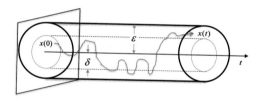

Stability is usually characterized in the sense of Lyapunov. To this end, Lyapunov functions (W) are required to prove the stability of a system. These need to be positive definite functions ($W(x) > 0$ for any $x \neq 0$). The Lyapunov stability theorem is as follows.

Theorem 2.1. *Let $x = 0$ be an equilibrium point for $\dot{x} = f(x, t)$, and let $D \subset \mathbb{R}^n$ be a domain containing $x = 0$. Let $W : D \to \mathbb{R}$ be a continuously differentiable function such that*

- $W(0) = 0$, and $W(x) > 0$ in $D - \{0\}$ (positive definite),
- $\dot{W}(x) < 0$ in D (negative semidefinite).

Then $x = 0$ is stable. Moreover, if

- $\dot{W}(x) < 0$ in $D - \{0\}$ (negative definite),

then $x = 0$ is asymptotically stable.

Proof. It can be found in [37]. $\square$

Note that the existence of Lyapunov functions is only a sufficient condition. That means if there is a Lyapunov function for a given system, then the system is stable; otherwise, it does not imply that the system is unstable. There is no general technique for constructing Lyapunov functions; in most of the cases, the construction of Lyapunov functions is a difficult task. For non-linear systems, local stability analysis can be performed on the linearized system at the equilibrium point, which writes as follows.

Theorem 2.2. *Let $x = 0$ be an equilibrium point for a nonlinear system $\dot{x} = f(x)$, where $f : D \to \mathbb{R}^n$ is continuously differentiable, and D is a neighborhood of the origin. Calculating the Jacobian matrix*

$$A := \left. \frac{\partial f(x)}{\partial x} \right|_{x=0}, \qquad (2.5)$$

the eigenvalues of the matrix A represented as λ can be computed. The origin is asymptotically stable if all eigenvalues of A have negative real parts ($\lambda < 0$). The origin is unstable if the real part of at least one eigenvalue of A is positive ($\lambda > 0$).

Proof. It can be found in [37]. □

2.4 **POPULATION MODELING**

Population modeling helps to describe dynamic changes of a population x such as birth, movement, interactions with other populations, death, and any other mechanism considered important. The main advantage of population modeling is that several processes that may be required to represent a mechanism can be summarized with a constant rate, reducing the complexity. A good bibliography for more detail of population modeling in the ecological field can be found in [38].

Through mathematical biology literature, one of the first and most well-known examples of population modeling is the Lotka–Volterra model, also known as the prey–predator model [39,38]. This model aimed to describe the dynamics of two species, one as a prey (x) and the other as a predator (z). As it was mentioned before, mathematical models have assumptions for which model predictions are valid. The Lotka–Volterra model has the following assumptions: *i*) food is not a limiting factor for the prey and can eat all the time; *ii*) the only food supply for the predator is the prey; *iii*) there is a constant prey production; and *iv*) the prey dies only due to the predator. The Lotka–Volterra model can be written as follows:

$$\frac{dx}{dt} = k_1 - k_2 x z, \qquad (2.6)$$

$$\frac{dz}{dt} = k_3 x z - k_4 z, \qquad (2.7)$$

where k_1 is the constant prey growth, k_2 represents the consumption rate of the prey by the predator, k_3 is the constant growth of the predator (note that the predator growth is directly proportional to the proportion xz of the pray and predator), and k_4 is the mortality rate of the predator. The negative sign means the loss of a population, whereas the positive sign implies the growth of a population.

2.7 SIMULATING LOTKA–VOLTERRA MODEL

```
% Lotka-Volterra Model Simulations
% Clearing window and memory
clc; clear all; close all;

% Simulation time
t0=200;

% Initial conditions
x_0=10;   % Pray
z_0=1;    % Predator

% Solving odes
[t,y]=ode45(@funodes,[0 t0],[x_0 z_0]);
% Changing variables for presentation
x=y(:,1); z=y(:,2);

% Plotting Results
figure(1);
plot(t,x,t,z,'LineWidth',2);
xlabel('Time','fontsize',20);
ylabel('Population number','fontsize',20);
legend({'Pray','Predator'},'FontSize',20);
hal=gca; set(hal,'LineWidth',2,'FontSize',20);
hold on

% The differential equations needs to be
written as a function either at the end of the
script or in another file with the same name.
function  dy = funodes(t,y)
% Changing of variables for presentation
x=y(1);
z=y(2);
% Model parameters
k1=1; k2=0.2; k3=0.05; k4=0.1;
% ODEs
dy = zeros(length(y),1);
dy(1) = k1*x -k2*x*z;
dy(2) = k3*x*z-k4*z;
end
```

Box 2.7 contains the Matlab code to simulate the dynamics between prey and predator and the simulation results that portray periodic solutions. It was observed that the population of the two species varied with the same period but somewhat out of phase. Note that there are several variations of the prey–predator model to construct more complex mechanisms. The prey–predator model and most of infectious disease models have as a principle the theory of chemical reactions [40].

The Lotka–Volterra model is the basis for the compartmental model of disease spread between hosts (epidemiological level), which can be expressed as follows:

$$\frac{dS}{dt} = -\frac{\beta}{N}SI, \tag{2.8}$$

$$\frac{dI}{dt} = \frac{\beta}{N}SI - \gamma I, \tag{2.9}$$

$$\frac{dR}{dt} = \gamma I, \tag{2.10}$$

where a population is divided into categories of susceptible (S), infected (I), and recovered (R). This is the reason why the model is known as the SIR model. The infection rate from susceptible to infected is considered with a constant rate β. Note that population is considered constant of size N. This assumes that birth and death processes are negligible mechanisms during the infections, for instance, flu infection. The parameter γ is the rate conversion from infected to susceptible. Given that $\frac{dS}{dt} + \frac{dI}{dt} + \frac{dR}{dt}$, system (2.8)–(2.10) has a constant population size at all times, that is, $S(t) + I(t) + R(t) = N$.

In fact, the SIR model can serve to describe not only between-hosts transmission but also within-host infections. A well-established and accepted mathematical model to represent viral dynamics is the target-cell-limited model; see details in Box 2.8. Although the target-cell-limited model is of simple structure, this has served to model several viral diseases; among the different viruses, there are HIV [41–43], hepatitis [9], influenza [8,44], ebola [45], dengue [46], and zika [47]. A detailed reference for modeling viral dynamics can be found in [42,48].

2.8 TARGET CELL MODEL

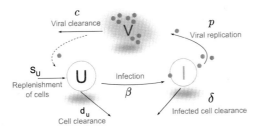

Target cells (U) are replenished with rate S_U and die with rate d_u. Virus (V) infects target cells (U) with rate β. Infected cells are cleared with rate δ. Once cells are productively infected (I), they release virus at rate p, and virus particles are cleared with rate c. Using ordinary differential equations (ODEs), the target cell model is considered as follows:

$$\frac{dU}{dt} = S_U - \beta U V - d_u U, \tag{2.11}$$

$$\frac{dI}{dt} = \beta U V - \delta I, \tag{2.12}$$

$$\frac{dV}{dt} = pI - cV. \tag{2.13}$$

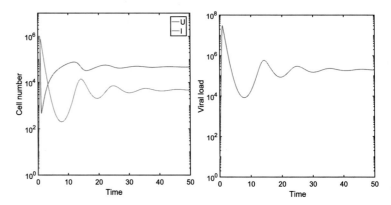

■ **FIGURE 2.1 Target-cell-limited model simulation.** Model initial conditions $U(0) = 10^6$, $I(0) = 0$, and $V(0) = 10$ were considered. Model parameters $d_u = 0.01$, $\beta = 10^{-7}$, $\delta = 2$, $p = 1000$, $c = 24$, and $s = U(0)d_u$ were assumed. This corresponds to the case of $R_0 > 1$.

Target cells can be either in a susceptible (U) or an infected state (I). Cells are replenished with a constant rate S_U and can die with rate d_u. Assuming the initial condition on susceptible cells is $U(0)$, the steady-state condition $S_U = U(0)d_u$ should be satisfied to guarantee homoeostasis in the absence of viral infection. Virus (V) infects susceptible cells with rate β. Note that the roles of the immune system are not modeled explicitly, but included in the rate of infected cells δ. Once cells are productively infected, they can release virions at rate p. Virus particles are cleared with rate c. In fact, the viral load will increase if $p \geq cV$, whereas it will decrease if $p \leq cV$ [42]. A numerical example of the target-cell-limited model is portrayed in Fig. 2.1, showing an exponential growth of infected cells and viral load. Consequently, attributed to the limitation of target cells, infected cells and viral load decreased. This simulation scenario represents the case of a chronic viral infection. Acute and chronic infections can be differentiated as follows.

Definition 2.31. **Acute infections** are caused by pathogens with fast growth rates promoting an infection that is cleared in a short period of time, for example, influenza.

Definition 2.32. **Chronic infections** are caused by pathogens with slow growth rates promoting an infection that is persistent for long-term periods. This occurs when the primary infection is not cleared by the adaptive immune response, for example, HIV.

Remark 2.5. Input (S_U) and natural death term $(d_u U)$ of susceptible cells are mechanisms particularly important because of the long-term dynamics in the case of chronic infections. The terms S_U and $d_u U$ may have a negligible impact in short dynamics, and thus for acute infections, Eq. (2.11) is usually considered as follows:

$$\frac{dU}{dt} = -\beta U V. \tag{2.14}$$

A key metric used in epidemiology is the reproductive number. Such concept helps determine whether or not an infectious disease can spread through a population.

Definition 2.33. Reproductive number (R_0) is the expected number of secondary individuals infected by an individual in its lifetime. If $R_0 \leq 1$, then the pathogen will be cleared from the population. Otherwise, if $R_0 > 1$, then the pathogen is able to invade the whole susceptible population.

Remark 2.6. The reproductive number for the target-cell-limited model (2.11)–(2.13) can be represented as follows:

$$R_0 = \frac{S_u \beta p}{d_u \delta c} = \frac{U_0 \beta p}{\delta c}. \tag{2.15}$$

The system of Eqs. (2.11)–(2.13) has two steady states, one where the host is free of disease,

$$E_1^* = \left(\frac{S_u}{d_u}, 0, 0 \right),$$

and the other steady state represents a chronic infection,

$$E_2^* = \left(\frac{S_u}{d_u} \frac{1}{R_0}, \frac{d_u c}{\beta p}(R_0 - 1), \frac{d_u}{\beta}(R_0 - 1) \right).$$

Theorem 2.3. *Assuming the target-cell-limited model (2.11)–(2.13), if $R_0 > 0$, then the equilibrium E_2^* is positive and globally asymptotically stable. If $R_0 < 0$, then the equilibrium point E_2^* is nonpositive, and the infection-free disease (E_1^*) is globally asymptotically stable.*

Proof. It can be found in [49]. □

Chapter 3

Model Parameter Estimation

CONTENTS

3.1 PARAMETER FITTING

Consider the following general system:

$$\dot{x} = f(x(t), \theta), \tag{3.1}$$

$$y = g(x(t), \theta), \tag{3.2}$$

$$\dot{\theta} = 0, \tag{3.3}$$

where $x(t) \in \mathbb{R}^m$ is the vector of state variables with initial condition $x(0) = x_0$, and $y \in \mathbb{R}^d$ is the output vector. For the parameter estimation problem, $\theta \in \mathbb{R}^q$ is a vector of constant parameters that can be estimated using the experimental data set $\hat{y} \in \mathbb{R}^d$. The initial condition x_0 is assumed to be known, which is not an equilibrium of the system and is independent of θ. Given system (3.1), the **inverse problem** consists on finding the parameter set θ based on the experimental data set $\hat{y}$. However, parameter estimation is a delicate process in mathematical modeling.

Parameter estimation can be online and offline. In the online approach, the model parameters are estimated when new data are available during the model operation. In the offline approach, all data are collected, and then the model parameters are estimated in a single process. Both approaches provide different results. For online estimation, model parameters can vary with time. This is attributed to the fact that online estimation is based on recursive algorithms that update the estimations every time a new measurement is available. For offline estimation, parameters are estimated and do not change due to a single optimization routine performed for estimation. The question which approach to use is based on the application. In the case of modeling infectious diseases or medical applications, usually measurements cannot be obtained in real-time but rather offline with infrequent sample periods and delays. Thus, this chapter is focused on offline parameter estimation.

Note that the analysis of biological experimental data has predominantly based on statistical methods. These approaches assist experimentalists to recognize differences and correlations, but in-depth interpretations of the underlying mechanisms are limited. With mathematical modeling approach, we can formulate different mechanisms in forms of mathematical relations. Consequently, parameter estimation procedures are performed to adjust model parameters against empirical data. Conducting parameter estimation requires familiarizing with different concepts of mathematics, optimization, programming language, and sometimes costly software toolboxes. Nevertheless, these technical problems should not prevent biologists and virologists from exploring their data potentials. Thus, in this chapter, we introduce an adaptable and state-of-the-art protocol for parameter estimation. Using ODEs to model viral infection, we adopt the target-cell-limited model presented, owing to its role as the core component of more than a hundred publications in virus modeling research, for example, influenza [7,8,33,50], HIV [51–56], Ebola [45,57], among several others.

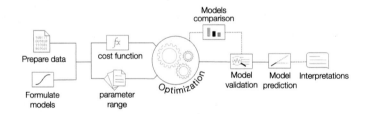

■ **FIGURE 3.1 Parameter estimation process in mathematical modeling**. Dashed lines indicate optional steps not presented in the scope of this chapter. Scheme was taken from [56].

Through the next sections of this chapter, the steps portrayed in Fig. 3.1 will be covered. Briefly, experimental data need to be prepared in standard formats. Model equations need to be defined with relevant components and corresponding model parameters. Based on that, a cost function that defines how matching the model and the data is written in the R programming language. The experimental data and the mathematical model can be then fed into an optimizer algorithm to find the best set of parameters that provide the best agreement between the model and data. The global optimization named differential evolution [58] is used here to adjust the model parameters. Setting conditions of the optimizer and the plausible range of the parameters of interests need to be defined. R codes in this chapter were taken from [56].

The proposed approach is applicable to influenza virus dynamics at different complexity levels; however, this can be adapted to any mathematical model and data set. For parameter estimation, additional R packages are needed for solving differential equations, such as deSolve [59], and for optimization based on the differential evolution algorithm, such as DEoptim [60]. These can be installed in R by running the following commands

```
install.packages("deSolve"); library("deSolve")
install.packages("DEoptim"); library("DEoptim")
# Remark: Anything that follows the character # is a comment and is
not processed in R. # Remark: The user should be aware that R is a
case-sensitive language, thus, check spelling carefully and whether the
letters are properly capitalized.
```

Virus dynamics as a study case. The development of mathematical models depends on several factors such as the data at hand, the hypothesis to be addressed, and the experience of the modeler. In this example, the target-cell-limited model for acute infections presented previously in Box 2.8 is considered. This reads as follows:

$$\dot{U} = -\beta U V, \tag{3.4}$$
$$\dot{I} = \beta U V - \delta I, \tag{3.5}$$
$$\dot{V} = p I - c V. \tag{3.6}$$

This mathematical model can be written in R format as follows:

```
myModel < - function(t,state,parameters) {
        with(as.list(c(state,parameters)),{
        # Fill in the equations
        dU = - myBeta*U*V
        dI = myBeta*U*V - myDelta*I
        dV = myP*I - myC*V
        # Outputs of the model
        list(c(dU, dI, dV))
        })
}
```

Note that if you see an error that says *non-numeric argument to binary operator*, it is probably that the name you gave to a model parameter already existed in R's default environments, for example, an expression such as Beta could lead to an error whereas myBeta would not. The user should be aware that the order of the return values in the deSolve model function is important [60]. They need to strictly follow the order listed by the model equations written above it.

The mathematical model is named as *myModel*. The three components U, I, and V, called states in modeling terms, are written in separate lines with leading letter d. The right-hand sides of the equations are the same as the model equations, except that the variable names are spelled out. The model initial conditions are the values of each variable in the model at day zero. Although the model parameters are in the log scale for parameter fitting, the model works in normal scales, and hence the initial conditions. Here the initial number of the target cells was approximated at 10^6 based on the experiment reports, and the initial number of infected cells at zero. The initial viral titers were set at 10 $TCID_{50}$. This value was chosen for this example at a value below the detection level of 50 $TCID_{50}$. Note that there is no conclusive method to define this number, whereas its value can affect the parameter accuracy [10]. Smoothing and extrapolating approaches seem to provide a reasonable estimate of the initial viral titers [10]. Thus, it is important to define initial conditions and parameters in the model for later use, including

```
myStates <- c(U = 10^6, I = 0, V = 10)
myParams <- c("myBeta","myDelta","myP","myC")  # list of parameters
myV <- "V"                        # variable that is observed in data
```

and the time scale for the model. This can be arbitrarily chosen, but it should cover the range of the observed time scale. Here the model output is recorded approximately every 15 min over ten days:

```
modelTime <- seq(from = 0, to = 10, by = 0.01)
```

3.2 EXPERIMENTAL DATA

Experimental data is a set of values able to help to interpret information within the values assigned to variables; such data sets should be collected in an experimentally repeatable manner. As a study case here, the mathematical model (3.4)–(3.6) was used to generate synthetic experimental data to resemble data set of influenza A virus infection with the viral dynamics and the sampling scheme resembling that of Toapanta and Ross [61]. To this end, host cells (10^6) and parameters values are assumed to fabricate influenza virus infection "data". Based on practical lab limitations, the viral load is assumed to be the only measurement, and thus viral titers were measured regularly (in $TCID_{50}$) at day 1, 2, 3, 5, 7, 9 post infection. At each time point, there are five replicates.

The data sets are presented in Table 3.1, which can also be downloaded in https://figshare.com/s/842b71f9c1bda1212a24. The first term in each row represents the sampling time, and the second is the viral load. Note that viral titers were already converted into log base 10 scales. The log scale viral titers were used not only because of its conventional usage in reporting viral load, but also because the log scale also implicitly assumes that viral load is log-normal distributed. This assumption simplifies the maximum likelihood problem to least squares and thus the use of the root-mean-square error (*RMSE*) as a cost function in the optimization [62].

Table 3.1 Fabricated viral load titers (in log base 10)

Day	Viral titers	Day	Viral titers	Day	Viral titers
1	2,36	3	5,21	7	3,67
1	3,24	3	5,44	7	3,23
1	2,70	3	5,56	7	3,36
1	2,57	3	5,60	7	2,85
1	2,38	3	6,12	7	3,76
2	4,87	5	4,97	9	1,62
2	4,42	5	4,55	9	1,81
2	5,05	5	4,66	9	2,00
2	3,77	5	5,06	9	1,70
2	4,16	5	4,80	9	2,23

The experimental data stored in Excel sheets are most often not ready for analysis in R. Comma-separated-values (.csv) is a universal format that can be read in any software.

The user can convert to .csv format by the following procedure:

1. Delete all irrelevant data (notes, comments, etc.) in the Excel sheet,
2. Name variables (columns) with computer-friendly format, i.e., no spaces or special characters, starting only with characters, not with numbers,
3. Choose Save As, and then Comma Separated in the file format field.

Note that to avoid unnecessary errors, CSV data files should be filled from the first row and first column, that is, the first row for variable names and the first column for data of the first variable. R can read a wide range of data types, even directly from Excel, but specific functions are needed. Double-checking data with decimal separators. Differences in locales of the operating system lead to wrong interpretation of numbers when saving as CSV. For example, a number stored in Excel as 3,141.2 would be potentially treated as two numbers, 3 and 141.2. This can be detected by visualizing the data in R or inspecting the CSV file directly to see how the data are interpreted. Numeric cells in Excel should be formatted with *"Use 1000 separator"* option disabled.

To read experimental data from .csv into R platform, run the following lines:

```
myData <- read.csv("/path/to/myData.csv")
myData                          # View data
```
 # Remark: On the Windows operating system, the path to the data file needs a special treatment in R, e.g., a file located in C:\Users\ABC\Downloads\myData.csv needs to be supplied to R as C:\Users\ABC\Downloads\myData.csv. This is because the backslash is also an escape character in R.

The data sets have two variables (each in a different column), including V (viral titers in log base 10) and the time (in days). For simplicity, the user can also input directly the data in R as follows:

```
# Write time in days
time <- c(1, 1, 1, 1, 1, 2, 2, 2, 2, 2, 3, 3, 3, 3, 3, 5, 5, 5, 5, 5, 7, 7, 7, 7, 7,
9, 9, 9, 9, 9)
# Write viral titres in log 10
V <- c(2.36, 3.23, 2.69, 2.57, 2.37, 4.87, 4.41, 5.04, 3.76, 4.16, 5.20,
5.43, 5.55, 5.60, 6.12, 4.96, 4.54, 4.65, 5.06, 4.80, 3.67, 3.22, 3.35, 2.84,
3.76, 1.61, 1.80, 2.00, 1.69, 2.22)
myData <- data.frame(time = time, V = V)
```

3.3 COST FUNCTION

To find the parameter set θ, it is necessary to use a measure function that evaluates the difference between the observations $\hat{y}$ and the model predictions by (3.1); this difference is usually called **residuals**. A widely applied measure function is the residual sum of squared errors (*RSS*). The smaller the *RSS*, the tighter fit of the model to the data. This can be written as follows:

$$RSS(\theta) = \frac{1}{2} \sum_{i=1}^{N} \left(y_i - \hat{y}_i \right)^2, \tag{3.7}$$

where $\hat{y}_i$ denotes the vector of experimental measurements at the time point i, and N is the total number of data points. The simulated output by the parameter set θ at the time point i is represented by y_i. Note that sometimes it is more convenient to use the root mean square (*RMS*), which gives an estimate of how far each measurement is from the mean. This can be written as follows:

$$RMS(\theta) = \left(\frac{2RSS}{N}\right)^{1/2}.$$ (3.8)

Let us now rewrite the least squares transforming the outputs and measurements in logarithmic scales as follows:

$$\chi^2(\theta) = \frac{1}{N}\sum_{i=1}^{N}\frac{\left(\log(y_i) - \log(\hat{y}_i)\right)^2}{\sigma_i^2},$$ (3.9)

where σ_i^2 is the corresponding standard deviation. The optimal parameter set θ^* can be estimated by

$$\hat{\theta} = \text{argmin}_\theta\left(\chi^2(\theta)\right),$$ (3.10)

where argmin stands for the argument of the minimum. When the output noise is normally distributed, $\epsilon \sim N(0, \sigma^2)$, χ^2 is equivalent to the maximum likelihood estimate (MLE) of θ, that is,

$$\chi^2(\theta) = \text{const} - 2\log(L(\theta)),$$ (3.11)

where $L(\theta)$ is the likelihood function [63]. The **likelihood** $L(\theta)$ function expresses how likely the outcome $\hat{y}$ is for different parameter sets θ. In other words, the likelihood is equal to the probability distribution of the observed outcomes $\hat{y}$ given the parameter set θ, that is, $L(\theta/\hat{y}) = P(\hat{y}/\theta)$. The maximum likelihood is a widely used frequentist approach for which the parameter set θ is fixed to the value that maximizes the likelihood function. To continue our parameter fitting example in R, the root mean square errors (*RMSE*) in log scales are considered to measure the magnitude of the difference between the output from the model (V) and the experimental data. This is formulated as follows:

$$RMSE(\theta) = \left(\frac{1}{N}\sum_{i=1}^{N}\left(\log(y_i) - \log(\hat{y}_i)\right)^2\right)^{1/2}.$$ (3.12)

Experimental data points $\hat{y}_i$ are represented in R language by *myData*, and the value y_i simulated from the model is represented by *ymodel*. In R language, the cost function can be written as follows:

```
myCostFn <- function(x)
      parms <- x[1:length(myParams)]
      names(parms) <- myParams
      ymodel <- ode(myStates, modelTime, myModel, 10^parms)
      yMatch <- ymodel[ymodel[,1] %in% myData$time, ]
      nm <- rle(myData$time)$lengths
      x <- myData[, myV]- rep(log10(yMatch[, myV]), times = nm)
      rmse <- sqrt(mean(x^2))
      return(rmse)
```

Note that because the model runs in a smoother time scale than the observed data, we calculated only the differences among the matched data points by the observed time. However, solving ODEs is not always easy when large combinations of the parameters are evaluated, some of which might not make sense at all. Thus, there might be cases where exact time points cannot be computed, but only their nearby neighbor. Matching these time points could lead to unexpected results. It is safer to truncate the numeric time points when matching the model time points and the data time points by replacing

```
ymodel[,1]%in% myData$time
with
as.character(ymodel[,1]) %in% as.character(myData$time)
```

3.4 OPTIMIZATION PROBLEM

Parameter procedures consist on minimizing the error between the model output and experimental data. Turn out that the minimization of the cost function (3.12) implies a nonlinear optimization problem, complex, with several variables, and multiple minima. This complexity can be tackled using evolutionary optimization algorithms such as the differential evolution (DE) algorithm [58].

The DE algorithm is a population-based optimization algorithm, where each individual in the population is an n-dimensional vector that represents a candidate solution to the problem. The individuals can be defined as:

$$x_{i,g} = [\chi_{i,g}^1, \chi_{i,g}^2, \ldots, \chi_{i,g}^n] \tag{3.13}$$

with $i = \{1, 2, \ldots, NP\}$, where $g = \{1, 2, \ldots, G\}$. $\chi_{i,g}^n$ are the elements of the ith individual in the generation g, n is the dimension of the problem, the variable G represents the maximum number of generations, and NP is the size of the population. The basic idea behind the DE algorithm is that two individuals mutually excluding $x_1 \neq x_2$ are picked randomly among population and its difference is scaled and added to a third individual $x_3 \notin \{x_1, x_2\}$ chosen randomly to create a new mutant vector v_i. The DE algorithm is explained with the following steps.

I) Initialization: the population is initialized within a bounded search space defined as $\chi_{\min} = \{\chi^1_{\min}, \ldots, \chi^n_{\min}\}$ and $\chi_{\max} = \{\chi^1_{\max}, \ldots, \chi^n_{\max}\}$, where $X_{\min}$ and $X_{\max}$ are the sets of lower and upper bounds for the elements of the individuals.

II) Mutation: different strategies exist for the mutation procedure. In this work, the following mutation strategies are considered [58]:

1. DE/best/1/

$$v_{i,g} = x_{\text{best},g} + F_{DE}(x_{r1,g} - x_{r2,g}). \tag{3.14}$$

2. DE/rand/1/

$$v_{i,g} = x_{r3,g} + F_{DE}(x_{r1,g} - x_{r2,g}), \tag{3.15}$$

where $v_{i,g}$ is the mutant vector, $x_{\text{best},g}$ is the individual with the best fit in the current generation, $x_{r1,g}, x_{r2,g}$, and $x_{r3,g}$ are different vectors chosen randomly among the population, and $F_{DE} \in [0, 1]$ is the scaling factor.

III) Crossover: a new trial vector is generated recombining the mutant $v_{i,g}$ and target $x_{i,g}$ vectors. The principal approaches of crossover are the binomial and exponential [58], described in Eqs. (3.16) and (3.17), respectively.

$$u^j_{i,g} = \begin{cases} v^j_{i,g} & \text{if} \quad r_{cj} \leq Cr \text{ or } j = j_{\text{rand}}, \\ x^j_{i,g} & \text{otherwise}, \end{cases} \tag{3.16}$$

where $j \in [1, n]$ is the jth component of $u_{i,g}, v_{i,g}$, and $x_{i,g}, r_{cj} \in [0, 1]$ is a random number taken from a uniform distribution, $Cr \in [0, 1]$ is a constant crossover rate, and $j_{\text{rand}} \in [1, n]$ is a random number taken from a uniform distribution.

$$u^j_{i,g} = \begin{cases} v^j_{i,g} & \text{if} \quad j \in A \cup B, \\ x^j_{i,g} & \text{otherwise}, \end{cases} \tag{3.17}$$

where $A := \{s_{DE}, \ldots, \min(n, s_{DE} + L_{DE} - 1)\}$, and $B := \{1, \ldots, s_{DE} + L_{DE} - n - 1\}$. The variable s_{DE} represents the starting element in a target vector to create a new trial vector recombining the L_{DE} elements donated by the mutant vector. Note that s_{DE} is an integer randomly taken from the interval $[1, n]$, and L_{DE} is drawn from the interval $[1, n]$ according to the following code:

```
1:  L_DE = 0
2:  while ((rand(0,1) ≤ Cr) and (L_DE < n)) do
3:      L_DE = L_DE + 1
4:  end while
```

where Cr is called the crossover rate. F_{DE}, NP, and Cr are the control parameters of the DE algorithm.

IV) Selection: the cost function is used to determinate which vector survives for the new generation $(g + 1)$, the mutant $v_{i,g}$ and target $x_{i,g}$ vectors are evaluated to find which vector has the best yield; the basic idea is as follows:

$$x_{i,g+1} = \begin{cases} u_{i,g} & \text{if } f\left(u_{i,g}\right) \leq f\left(x_{i,g}\right), \\ x_{i,g} & \text{if } f\left(x_{i,g}\right) < f\left(u_{i,g}\right), \end{cases} \tag{3.18}$$

where $f(\cdot)$ is the evaluation of the objective function.

Note that some characteristics from a generation can be transferred to the next one in the DE/best/ variants. In this way, the members of the next generation can have certain properties of the current leader, reducing the space of search. Otherwise, the next generation is randomly generated favoring the space of search (DE/rand/).

Running the Optimizing Toolbox

Based on the literature, we define ranges of plausible parameter values. For example, the elimination half-life $(t_{1/2})$ of influenza is unknown; however, this cannot be considered either in seconds or in a century time scale. Converting this time into the clearance rate c is done by the formula $t_{1/2} = \ln(2)/c$. Note that the boundaries are transformed to log base 10 scale and that the numbers are subject to one's own expertise. Readers can use this formula to compute our assumptions on the rates of infected cell death, viral replication, and viral clearance based on the boundaries of the three parameters δ, p, and c:

```
lower = log10(c(1e-7, 1e-2, 1e+0, 1e-1))
upper = log10(c(1e-3, 1e+2, 1e+2, 1e+2))
```

Some model parameters have analytical meanings but no equivalent experimental values. For example, the parameter β represents the reaction rate between the virus and the target cells, which depends on the concentration of the virus and the number of target cells. In such cases, the parameter range was defined to cover several orders of magnitude.

Standard optimizer settings are often insufficient for complex models. The optimizer can be forced to work more exhaustively by increasing the number of trials (combination of *itermax* and *steptol*) and decreasing the relative tolerance named as *reltol* (measurement of the error relative to the size of each solution component).

```
myOptions <- DEoptim.control(itermax = 10000, steptol = 50, reltol =
1e-8)
```

At this point, we are ready to fit the model by calling the optimizer with the inputs including the cost function, the lower and upper bound, and the options as follows:

```
fit <- do.call("DEoptim", list(myCostFn, lower, upper, myOptions))
```

The simulation time took approximately five minutes on an Intel Core i7, 8 Gb RAM computer. The algorithm evaluates different combinations of the parameters in the provided ranges, comparing them by the RMSE. The process is repeated until the algorithm fails to find a combination of parameters that has a significant improvement compared to the current parameter set.

If the user observes no changes in the cost function output (*bestval*) after several iterations or even after reaching the maximum number of iterations (*itermax*), then this means that the optimization failed. The user can try to adjust the parameter boundaries to a more probable region based on the literature and the meaning of the parameter. As noted by the Soetaert et al. [60], if the user observes the error

```
DLSODA- Above warning has been issued xx times,
```

then this means that the ODEs solver could not proceed with the current set of parameters. The most likely reason is that the parameter range was too wide, leading to extreme values to be evaluated. This can be avoided by narrowing down the parameter ranges to a more plausible region. Visualization of the results can be done with the following commands:

```
(bestPar <- fit$optim$bestmem) # parameter estimates in log 10 scale
names(bestPar) <- myParams
out <- ode(myStates, modelTime, myModel, 10^bestPar)
plot(out[, "time"], log10(out[, "V"]), type = "l") # in log 10
```

points(myData$time, myData$V) # superimpose observed data points
Remark: The following figure is printed by R

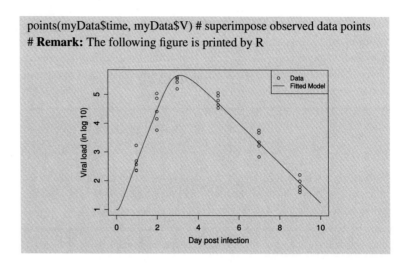

Note that the differential evolution algorithm can be sped up considerably
with parallel mode enabled with the option such as

myOptions <- DEoptim.control(parallelType = 1, packages = c("deSolve"), parVar = c("myModel","myStates", "lower", "upper", "modelTime", "myData"), itermax = 10000, steptol = 50, reltol = 1e-8)

However, successful application of this method cannot be guaranteed for all
calculations. Certain understanding about parallelization computing in R is
needed to avoid miscalculations, for example, by providing wrong data or
variables for calculation.

3.5 **IDENTIFIABILITY**

Redundant parameterization in mathematical models provides structural
identifiability problems due to an insufficient mapping in the function g
of the internal model states x to the output measurements y. Ambiguous
parameters $\theta_{sub} \subset \theta$ may be varied without changing the output y resulting
in constant values for χ^2. In biology, a large variability is presented from
one host to another, limiting the prediction value of mathematical models
and estimated parameters [64]. Identifiability concepts introduced in this
section are taken from [32].

Definition 3.1. Identifiability. System (3.1) is identifiable if θ can be
uniquely determined from the measurable output $y(t)$; otherwise, the system is unidentifiable.

A system that is controllable and observable has strong connections among input, states, and output variables. Such a strong connectivity may indicate that the system is identifiable [32]. Furthermore, [65] introduced the differences between global and local identifiability:

Definition 3.2. Global identifiability: Equation system (3.1) is globally identifiable if, for any two parameters vectors θ_1 and θ_2 in the parameter space $\Theta \subseteq \mathbb{R}^q$, $y(\theta_1) = y(\theta_2)$ if and only if $\theta_1 = \theta_2$.

Note that the concept of global identifiability is too strong for practical use. For biological applications, the concept of local identifiability is more appropriate.

Definition 3.3. Local identifiability: System (3.1) is locally identifiable if, for any θ within an open neighborhood of some point θ^* in the parameter space $\Theta \subseteq \mathbb{R}^q$, $y(\theta_1) = y(\theta_2)$ if and only if $\theta_1 = \theta_2$.

These two definitions imply a one-to-one mapping between the parameters and output. More restrictive conditions for differential equations can be stated when the initial condition is given [66].

Definition 3.4. Local strong identifiability (x_0-identifiability): The system structure is considered locally strongly identifiable if, for a given initial state $x_0 = x(t_0)$, which is independent of θ and not an equilibrium point, there exists an open set Θ^0 within the parameter space Θ such that, for any two different parameter vectors $\theta_1, \theta_2 \in \Theta^0$, the solutions $x(t, \theta)$ exist on $[t_0, t_0 + \epsilon]$ ($t_0 < \epsilon \leq t_1 - t_0$) for both θ_1 and θ_2, and $y(\theta_1) \neq y(\theta_2)$ on $[t_0, t_0 + \epsilon]$ for some $\epsilon \in (0, \infty)$.

Furthermore, identifiability definitions based on a one-to-one mapping between system parameters and system output was proposed by [65] developing the definition of identifiability based on the algebraic equations:

Definition 3.5. Algebraic identifiability: Based on algebraic equations of the system state and output, the system structure is considered to be algebraically identifiable if a real analytic function

$$\phi = \phi(\theta, y, \dot{y}, \ldots, y^{(k)}), \quad \phi \in \mathbb{R}^q,$$

can be constructed after a finite number of steps of algebraic calculations such that $\phi = 0$ and the matrix $\frac{\partial \phi}{\partial \theta}$ is nonsingular in the time range of interest $[t_0, t_1]$ for any (θ, x_0) in an open dense subset of $\Theta \times M$, where k is a positive integer, M is an open set of initial states, and $y, \dot{y}, \ldots, y^{(k)}$ are the derivatives of y.

Structural identifiability techniques verify identifiability by exploring the model structure. Xia et al. [64] proposed a method based on the implicit function theorem. This reads as follows.

Theorem 3.1. *Let* $\phi : \mathbb{R}^{d+q} \to \mathbb{R}^q$ *be a function of the model parameter* $\theta \in \mathbb{R}^q$, *system output* $y \in \mathbb{R}^d$, *and its derivatives, that is,*

$$\phi = \phi(\theta, y, \dot{y}, \ldots, y^{(k)}),$$

where k is a nonnegative integer. Assume that ϕ has continuous partial derivatives with respect to θ. A system is said to be locally identifiable at θ_* *if there exists a point* $(\theta_*, y_*, \dot{y}_*, \ldots, \dot{y}_*^{(k)}) \in \mathbb{R}^{d+q}$ *such that*

$$\phi(\theta_*, y_*, \dot{y}_*, \ldots, y_*^{(k)}) = 0$$

and the identification matrix

$$\left. \frac{\partial \phi}{\partial \theta} \right|_{\theta=\theta_*} \tag{3.19}$$

is nonsingular. Using the Taylor expansion of ϕ at θ_*,

$$\phi \approx \phi(\theta_*) + (\theta - \theta_*) \left. \frac{\partial \phi}{\partial \theta} \right|_{\theta=\theta_*},$$

since $\phi(\theta_*) = 0$ and $[\frac{\partial \phi}{\partial \theta}|_{\theta=\theta_*}]^{-1}$ exists, a unique solution of θ can be found, and the system is locally identifiable at θ_*.

Proof. The proof is provided in [64]. □

3.9 STRUCTURAL IDENTIFIABILITY IN THE TARGET CELL LIMITED MODEL

A system is structurally identifiable if the identification matrix (3.19) is nonsingular. This methodology was applied by Miao et al. [32] to the target-cell-limited model described in Box 2.8, that is, Eqs. (3.4)–(3.6). For the parameter set $\theta = [\beta, \delta, p, c]$, assuming that the viral load V is the only measurement available, system (3.4)–(3.6) is algebraically identifiable if there exists a function ϕ such that $det[\frac{\partial \phi}{\partial \theta}] \neq 0$ and $\phi(\theta, V, \dot{V}, \ldots, V^k) = 0$ on $[0, t]$. Using high-order derivatives in system (3.4)–(3.6), Miao et al. [32] obtained the following function:

$$\phi_1 = (V^{-1}\dot{V} - \beta V)\left(\ddot{V} + \delta c V + (\delta + c)\dot{V}\right) - (\delta + c)\dddot{V} - c\delta\dot{V}. \tag{3.20}$$

Note that the identification equation (3.20) does not depend on the unobservable states. Thus further identifiability studies can be performed.

Remark 3.1. The previous works [32] and [51] showed that the identification equation (3.20) does not contain the parameter p, and therefore this parameter is not structurally identifiable. However, [51] remarked that the parameter p can be identified if the initial conditions are known.

Definition 3.6. Practical identifiability. A system that is algebraically identifiable may still be practically nonidentifiable if the amount and quality of the data are insufficient and manifest large variability.

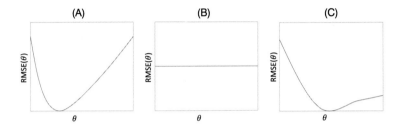

■ **FIGURE 3.2 Profile Likelihood**. Continuous lines present the profile likelihood versus parameters. Panel (A) presents an identifiable parameter. Panel (B) shows a parameter with structural identifiability problems. Panel (C) shows a parameter with practical identifiability problems.

The novel approach proposed by Raue et al. [63] exploits the profile likelihood to determine both structural and practical nonidentifiability. The idea of this approach is to explore each parameter in the direction of the least squares $(\chi^2(\theta))$. The profiling of model parameters is used then to inform the identifiability of parameters or to calculate the parameter confidence interval. For each parameter, a sequence extending both sides of the estimated parameter to the boundaries is generated. For practical identifiability, it is necessary to identify in which directions χ^2 flattens out; see Fig. 3.2C. Because of the complexity of the model and the scarcity of experimental data, it might be impossible to identify the parameters with less uncertainty [63]. For each value, optimize by keeping the parameter value fixed while readjusting the other parameters. For structural identifiability, the profile does not flatten but maintains a constant *RMSE* value while the parameter is varied; see Fig. 3.2B. A parameter can be identified when the profile likelihood presents a concave form as shown in Fig. 3.2A. The profiling function can be programmed in R as follows:

```
myProfile <- function(lower, upper, bestPar) {
    pro.ll <- NULL
    for (v in 1:length(bestPar)) {
    # Create parameter sequence
```

```
    tmpl <- seq(lower[v], bestPar[[v]], length.out = 100)
    tmpl <- tmpl[order(tmpl, decreasing = TRUE)[cumsum(1:13)]]
    tmpr <- seq(bestPar[[v]], upper[v], length.out = 100)
    tmpr <- tmpr[cumsum(1:13)]
    pars <- sort(unique(c(lower[v], tmpl, bestPar[[v]], tmpr, upper[v])))
    ppl <- NULL
    # Run optimization for each, and record the parameters and RMSE
      for (p in pars) {
          DEargs <- list(myCostFn, replace(lower, v, p), replace(upper, v,
p), myOptions)
          fit <- do.call("DEoptim", DEargs)
          ppl <- c(ppl, fit$optim$bestval)
          }
      pro.ll[[v]] <- cbind(pars, ppl)
      }
    return(pro.ll)
}
```

For efficiency, the parameter sequences were chosen such that they are dense close to the estimated values and are sparse toward the boundaries. The profiling can now be performed simply by executing the following function:

```
outProfiles <- myProfile(lower, upper, bestPar)
```

For each parameter, this function calculates 26 discrete points to form the likelihood profile. In total, more than a hundred optimizations will be run, which can take hours or even days depending on the power of the computer. Note that many parts of the codes can be sped up by vectorizing and parallelizing the calculations, for example, using mcapply, snow package. Different implementations are needed for Windows compared to Linux and Mac. Here, using base R codes, the calculations can be applied to all operating systems. Plotting the profile of the first parameters can be done by executing

```
par(mfrow = c(2, 2))
sapply(1:4, function(x) plot(outProfiles[[x]], xlab = myParams[x], ylab
= 'RMSE'))
```

3.6 **BOOTSTRAPPING PARAMETERS**

Bootstrapping is a statistical method for assigning measures of accuracy such as confident intervals to parameter estimates [67]. Here a nonparametric approach using Monte Carlo resampling is employed. First, we resample the data with replacement to have a sample of equal size to the original data. The parameters are then estimated from the resampling. The procedure is repeated many times to obtain the bootstrap distributions of the parameters.

```
myBoot <- function(numboot = 1000, numpar = 4) {
    # numpar: number of parameters in the model
    # numboot: number of bootstrap samples
    results <- matrix(NA, numboot, numpar)
    original <- myData
    sampling <- function(x) sample(original$V[original$time==x],
    length(original$V[original$time==x]), replace = 1)
        for (i in 1:numboot) {
            message("Bootstrapping sample ", i)
            tmp <- sapply(unique(original$time), sampling)
            myData <- cbind(original$time, as.vector(tmp))
            DEarguments <- list(myCostFn, lower, upper, myOptions)
            fit <- do.call("DEoptim", DEarguments)
            results[i, ] <- fit$optim$bestmem
        }
    results <- as.data.frame(results)
    colnames(results) <- myParams
    return(results)
}
```

3.7 SOURCES OF ERRORS AND LIMITATIONS

Mechanistic models aim to mimic biological mechanisms that embody some essential and exciting aspects of a particular disease. These can either serve as pedagogical tools to understand and predict disease progression or act as the objects of further experiments. However, mathematical models at heart are incomplete, and the same process can be modeled differently, if not competitively. Inherently, a model should not be accepted or proven as a correct one but should only be seen as the least invalid model among the alternatives. This selection process can be done, for example, by fitting the models to experimental data and comparing the model goodness-of-fit criterion; see, for example, [32,33,50].

Assuming that there exists a model that represents properly the problem at hand, model fitting to experimental data is still subject to a number of factors that can distort parameter estimates. For instance, mathematical models are often nonlinear in structure, and their parameters can be strongly correlated, posing troublesome issues to parameter estimation procedures, for example, parameter identifiability or sensitivity. These problems aggravate as experimental data are often sparse, scarce, and varying by orders of magnitude [32,51,63,64]. Consequently, accurate estimation of biological parameters is inherently problematic, if not impossible.

Nevertheless, previous issues are not de facto worrisome barriers if the primary purpose is to evaluate working hypothesis. In this case, finding a working set or a plausible range of parameter values can be deemed sufficient. However, doubts arise when mathematical models are used explicitly to extract biological meaning of model parameters, which is more evident and likely to happen in those applications where the prediction aspect is inferior. For example, modeling research on acute diseases, such as influenza virus infection [7], has been pursuing the quantification of parameters rather than prediction. Additionally, the conventional practice of making use of the formerly estimated parameters in mathematical modeling could propagate the potential inaccurate parameters to the subsequent work and weaken the validity of parameter estimates and consequently the model merit. Several approaches have been offered to provide assessments and understanding on the soundness of the estimated parameters [63]. However, in fact, the booming of mathematical models in biological and medical research over the last years has been accompanied with a disproportionate amount of assessments on validity of parameters. This quantitative recapitulation of biological parameters without doubt has raised concerns in the scientific community, provoking questions on the appropriateness of the approach.

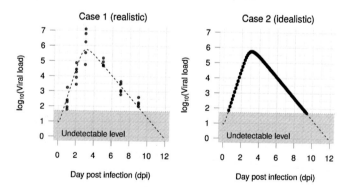

■ **FIGURE 3.3 Fabricated data for parameter estimation process**. The dotted curve is the true kinetics, dots are data points [10].

To keep modeling biological systems from being mistrusted, an important step is to recognize potential sources of error and to assess them in publications. This practice will not only increase the parameter credibility but also save future research from repeating the same mistakes. In this section, numerical simulations of a popular mathematical model in viral infection research from [10] are used to exemplify various sources of error that exist even when a correct and structurally identifiable model is used to fit to designated data. Using fabricated data presented in Fig. 3.3, different angles of the limitations in quantifying biological parameters in mathematical modeling are exemplified.

Simulation scenarios were differentiated in number and spacing of sampling time points, number of replicates per time point, and whether a certain experimental setting is known. The reference parameters were chosen to have synthetic data that closely resemble typical experimental observations. In particular, influenza A virus (IAV) infection kinetics comprises a quick viral replication and a peak around the 2nd and 3rd day post infection (dpi) [7]. The viral load stays in a plateau shortly before it declines to an undetectable level below the limit of detection (LoD) at days 7–12 onward [7,8]. The half-life of infected cells ($\delta = 1.6$) and viral clearance rate ($c = 3.7$) are assumed. The remaining two parameters were chosen such that the viral load exponentially increases and peaks at the 3rd dpi then decline to the undetectable level at the 9th dpi, that is, $\beta = 10^{-5}$ and $p = 5$. Unless stated otherwise, the initial number of uninfected cells, infected cells, and virus inoculum are assumed known as $U_0 = 10^6$, $I_0 = 0$, and $V_0 = 10$, respectively. To reflect the LoD, generated data points that are smaller than 50 pfu/ml were removed. The measurement error is assumed to be normally distributed on log base ten, that is, $\sim N(0, 0.25)$, unless otherwise stated. Fig. 3.3 visualizes the two examples of the generated data that mimic a realistic and an idealistic situation.

Choices of Initial Conditions in Skew Parameter Estimates

Generally, the model initial conditions can be included in parameter estimation procedures as unknown parameters. However, this can be intractable as more nuisance parameters are involved in the estimation routine. An appealing approach is to fix initial conditions and estimate only the model parameters; see, for example, [7]. For the target cell model, it is reasonable to assume that the number of infected cells at the beginning of infection is zero. However, the initial number of target cells and inoculum are experiment-dependent and can only be roughly approximated; for example, initial inoculum had been chosen as half of LoD [68].

To isolate the effect of initial conditions on parameter estimates, an idealistic scenario was generated (Fig. 3.3, Case 1), and the model was fitted with the DE algorithm. When the correct initial conditions of the model were used, the DE algorithm returned the exact values of the parameters. Afterward, the initial value of each component in the model was varied and reestimated the parameters and the corresponding profile likelihood. Because the data and the estimation algorithm did not changed, any error emerging in the parameter estimates can be attributed to the changes in the initial conditions. The tested values of the initial conditions were chosen to reflect biological range and its uncertainty in practice, that is, the initial viral load was under the LoD while the number of the initial target cells was about 10^7 (Fig. 3.3).

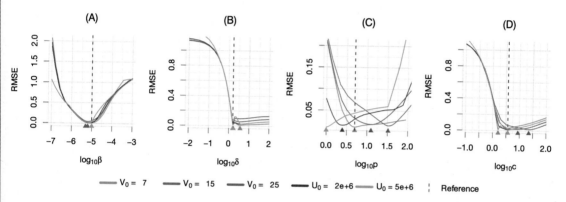

■ **FIGURE 3.4 Parameters profiles in different initial conditions**. The vertical dashed lines indicate the reference values, the triangles are the points estimated for the corresponding initial condition. The data include thirty replicates per time point with minimal measurement error; samples are collected in every three hours during 12 days of the experiment [10].

Simulation results show that any deviations, even those that seem negligible, result in erroneous parameter estimates (Fig. 3.4). Arbitrary choices of the initial viral load lead to very different estimates of essential viral infection kinetics parameters, for example, both the replication rate p and the viral clearance c were estimated six times higher than the true value when fixing the initial viral load at half LoD. Furthermore, subjective choices of model initial conditions result in flatness in the profile likelihood and shift the minimum in the parameters space. Note that the cost function values, that is, the root mean squared error (RMSE) in this case, of different parameter sets are still almost identical. Smoothing methods provide efficient ways to illustrate data dynamics; more importantly, these are free from the straitjacket of distributional assumptions and model structure. As such, smoothing techniques can be potentially useful in providing estimates for missing data points and assisting the parameter estimation [10].

Measurement Error Impairs the Parameter Estimates

The previous scenario assumed virtually no measurement error, whilst experimental data in practice can often be seen to vary in one to three orders of magnitude. In this context, varying the imposed measurement error of synthetic data and profiling the parameters can provide useful hints on the possible influence of measurement error on parameter estimates. Using the correct initial conditions, an idealistic data set with a large number of replicates and dense sampling time was developed to isolate the effect of measurement error on the accuracy of parameter estimates. Parameter estimation was done with the DE algorithm as it had recovered the exact parameters when there were minimal measurement errors (Fig. 3.5). As expected, the results show that the estimates drift away from the reference value as the measurement errors increase (Fig. 3.6). More importantly, increasing the measurement error brings in more flatness to parameter profiles, implying difficulties in estimating parameter values with accuracy.

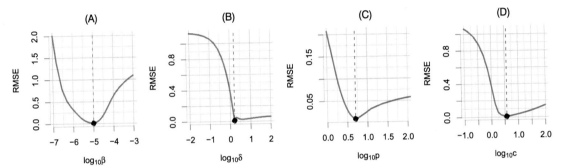

■ **FIGURE 3.5 Parameters profiles in an idealistic scenario**. The vertical dashed lines indicate the reference values, the solid circles are the estimated values from DE algorithm. The data include thirty replicates per time point with minimal measurement error, i.e., $N(0, 0.01)$ on log base ten, samples are collected in every three hours during 12 days of the experiment [10].

Data Asynchrony Generates Overlooked Measurement Error

Experimental data are usually taken for granted as it is observed timely. However, due to different reasons (e.g., host factors, experimental settings, and measurement techniques), the observed data are likely to become *asynchronous*. In such situations, measurements recorded at one time point are at best an aggregate measure of the neighbor period. Let t denote the real time point when an observed value actually happens; the time point will be instead recorded as $t^* = t + v$, $v \sim N(0, \sigma_t^2)$, where v is an unknown time shift. This shifting can be attributed to the differences in subject-specific responding time to a stimulus and its strength, for example, infection and the virulence of a virus strain. Furthermore, it is practically not feasible to infect all experimental subjects at the same time or to harvest all required replicates at once. For instance, subject one (Fig. 3.7A) for unknown reasons has ten hours delay in response to infection, but it is harvested at the first day in the *experimental time scale*. This shifting underestimates the viral load on the first day by one order of magnitude. By and large, the asynchrony leads to data with significant variations even in the absence of measurement error (Fig. 3.7B). The scope of this problem is largely unexplored in mathematical modeling research [10].

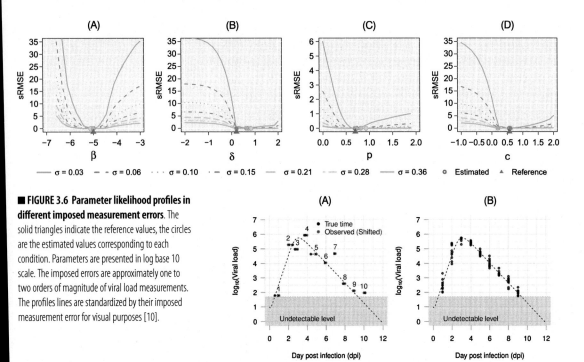

■ **FIGURE 3.6 Parameter likelihood profiles in different imposed measurement errors**. The solid triangles indicate the reference values, the circles are the estimated values corresponding to each condition. Parameters are presented in log base 10 scale. The imposed errors are approximately one to two orders of magnitude of viral load measurements. The profiles lines are standardized by their imposed measurement error for visual purposes [10].

■ **FIGURE 3.7 Examples of asynchrony problem**. The dotted curve is the true kinetics, dots are data points. (A) Ten subjects with measurement error and asynchrony; red (dark gray in print version) points are the observed data; black points are the actual time the measurement should have represented; arrows indicate directions of time shift. (B) Generated asynchrony data in the absence of measurement error.

Sparse and Scarce Data Aggravate the Parameter Estimates

Data sampling in experimental settings is laborious and costly, and, consequently, experimental data are usually sparse and scarce. In influenza infection, for example, the viral load had been measured in time intervals as short as eight hours up to four days apart [61]. In each time point, there could be only a single replicate [69,70] or up to sixteen replicates in rare cases [61]. In this context, investigating the impact of different data sampling schemes on parameter estimates provides further angles on how prone to error the parameter values can be in practice.

To explore the impact of the sampling time scheme in more practical settings, scenarios in which the data are collected at various time intervals were generated, assuming ten replicates per time point with realistic measurement errors. For each scenario, the estimation using the DE algorithm was repeated on a thousand generated datasets to calculate bias and variance.

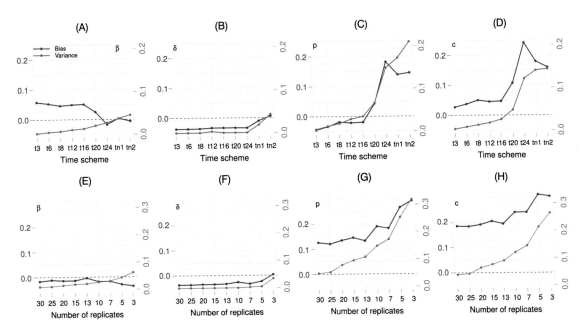

Unsurprisingly, the parameter estimate variance increases as the sampling time gets sparser (Fig. 3.8A–D). Bias-variance of both parameters p and c escalates from 16-h time scheme to 20-h time scheme. It can be observed in Fig. 3.8 that more precision is obtained by increasing the sampling time frequency, but little is gained in parameter accuracy. The practical experiment sampling time tn1 (6 time points and 5 replications) and tn2 (sampling every day but missing the peak day and last time point) manifest the highest variance (Fig. 3.8).

Analysis of the number of replicates per time point was also examined in a similar fashion. The every day sampling scheme (t24) was chosen to see if the estimates could be improved by changing the number of replicates per time point. As expected, the variance of most parameters increases when the number of measurements per time point is reduced, as well as the bias (Fig. 3.8E–H). There is no much gain in parameter accuracy as the number of replicates per time point roughly doubles from 13 to 30, and even less for the parameters β and δ.

Experimental design exhibits a strong influence on the magnitude of the bias, that is, varying either the sampling time or the number of replicates per time point yields steep changes in bias magnitude. It seems that experimental design rewards more accuracy in parameter estimates by sampling more frequently rather than sampling more extensively (Fig. 3.8). It is worth noting that the examples thus far have shown that some model parameters

■ **FIGURE 3.8 Sampling time scheme and parameter bias-variance.** **Top**: t3 to t24 indicate regular sampling time in every 3 to 24 hours respectively; tn1 indicates sampling time similar to that of [50]; tn2 is t24 without measurement at the peak day (3rd) and the endpoint (8th). Ten replicates were generated per time point. Each scenario (point) is the result of 1000 simulations. **Bottom**: The every day sampling scheme (t24) is used.

were always estimated with larger error than the others given the same context (Fig. 3.8). This underlines the discrepancy of parameter accuracy level in a model, that is, differences of parameter roles in a model lead to different levels of parameter accuracy.

Performance by Different Parameter Estimation Algorithms

Thus far the DE algorithm has shown that it can recover the exact parameters in an idealistic case. However, using a wrong assumption for the model or poor experimental data can both drive the algorithm to the wrong estimates. With the progress in parameter estimation techniques these days, it would be intriguing to know whether there is an algorithm that can recover more accurate estimates given a realistic experimental data set. It has been shown that there is not a definitive algorithm recommended for all situations [10]. To this end, the state-of-the-art algorithms are tested to estimate the four model parameters in a practical context (Fig. 3.3, Case 1). It is assumed that the initial conditions are known to make the parameters structurally identifiable [32]. The tested algorithms include a box-constraint local estimation algorithm (L-BFGS-B), a global stochastic algorithm, that is, differential evolution (DE) algorithm, and a Bayesian sampling approach, that is, adaptive Metropolis–Hasting (MH) algorithm. Maximum likelihood estimation (MLE) and least square (LS) estimation were considered where relevant to reflect their usage in practice. To avoid imposing subjective bias in favor of one algorithm, the algorithms were supplied with their recommended settings.

Simulation results show that the algorithms return relatively similar, whilst inaccurate point estimates of the parameters, for example, the two parameters p and c, are distanced by one order of magnitude from the correct ones (Fig. 3.9). Interval estimates of the three algorithms cover similar ranges that spread up to two orders of magnitude. Good interval estimates can be obtained only for the parameter δ. Note that the bootstrapping samples exhibit multimodal curves, implying that there is at least one outlier in the used data. Given the same advantages, the intensive estimation processes do not reward better estimates, that is, the L-BFGS-B converged in a few minutes, the MH took twenty minutes to converge to a stable distribution, and the bootstrapping with the DE algorithm took a day on a high-performance computer with parallel mode enabled.

The estimations were conducted in very favorable conditions, which are unlikely to occur in practice, that is, a correct model of the observed data, known initial condition values, real parameter values are in the ranges of

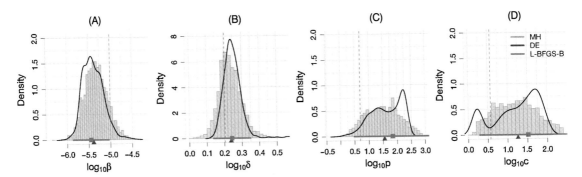

■ **FIGURE 3.9 Parameter estimates using different algorithms**. The data are collected similar to [50] at day 1, 2, 3, 5, 7, 9, five replicates per time point. **MH**: Metropolis–Hasting estimates are presented as blue (light gray vertical rectangles in print version) histograms of posterior samples. **L-BFGS-B**: local optimization and confidence intervals estimated with approximate Fisher information are presented with red (mid gray horizontal line with a rectangle in print version) square points and bars. **DE**: point estimates of the global optimization algorithm DE and weighted bootstrapping samples are presented with black triangles and curves. The vertical red (mid gray in print version) dashed line is the reference value. Figure taken from [10].

searching parameter spaces. Even with those advantages, the parameter values are risky to be interpreted, for example, the viral replication rate p can be recognized as much as one hundred copies per day with strong supports from the interval estimates of different algorithms. Certainly, expecting more accurate estimates in practice is rather optimistic, especially, when most often none of the above advantages exist in practice. Note that the same settings for the DE algorithm has been shown to be able to recover the exact parameters in an idealistic data set and the MH algorithm had converged to a stable distribution. In addition, a test for different combinations of the DE algorithm parameters was conducted showing that tuning these DE algorithm parameters did not improve the RMSE. Thus it can be inferred that the main component that has driven algorithms to the wrong estimates was a poor experimental setting. This example also suggests an influential role of outliers in estimating mathematical model parameters. Sources of outliers data could be merely originated from measurement errors or, in fact, come from a different generating mechanism.

3.8 **CONCLUDING REMARKS**

Mathematical modeling research comprises several disciplines with mutual concepts, approaches, and techniques. Significant concerns about the validity of parameter values had emerged when a majority of modeling publications failed to mention the risks of errors in the parameter values while trying to extract biological meaning from it. Throughout this chapter, by presenting numerical examples and addressing their implications we explored how the parameter estimates are error-prone in everyday practices.

It is a catch-22 situation where the way we express a parameter in the model governs its own accuracy. The results show that given the same context, there are model parameters that can be estimated with less error while the others cannot. This reflects the reality that the corresponding component of a parameter in the model is less important in the dynamics. To provide assessments in this aspect, we can head for a global sensitivity analysis approach. However, in biological systems, we can expect joint effects of parameters on the system output. This leads to situations in which an inaccurate set of parameters generates the same dynamics as the correct one. In this case, the estimated parameter values need to be taken with caution as they are a possible solution in a pool of model solutions. Altogether, in the worst case, the model structure alone already hinders the parameter accuracy, making the interpretation of the parameter values questionable. Reporting assessments of model parameters sensitivity, parameter correlation, and identifiability (see, e.g., [45]) can avoid the model results to be overinterpreted.

Leaving out the validity of a model to represent a certain phenomenon, parameter estimation of a correct model with known parameters and fabricated data exhibited several shortcomings. Each algorithm has per se a number of parameters and configurations that are required to be specified by the user [10]. The way we choose these settings may directly affect the estimation results. For instance, algorithms might be trapped in a local optimum, and this can be tested by tuning the algorithm. Although assessing this aspect is beyond the scope of the work proposed here, simulation results support that the recommended setting is among the best ones for the DE algorithm. In addition, testing different algorithms for a study case can help to find out if one algorithm was trapped in a local minimum. However, even in ideal conditions, our results show that different algorithms returned erroneous estimates. Note that the model parameters in the tested situation were set to be theoretically identifiable. This result poses a big question mark on the validity of mathematical model parameters in practice, where the models are more complex, but the known model conditions and experimental data are limited. More important, there is not a conclusive method for parameter estimation.

In experimental settings, several sources of errors exist that are difficult to overcome. In the target cell model, for example, it is not an easy task for experimentalists to define the exact initial conditions even with the advance of laboratory devices. In contrast, mechanistic models work with a relatively high precision level. This is a straitjacket for all modelers. Working with assumed model initial conditions might be the first choice and could be a valid one if the aim is not biological parameter quantification. This approach reduces the number of parameters to estimate in the shortage of data and may hold up in model prediction contexts. However, numerical results show that the estimates went wrong with seemingly reasonable choices of initial conditions in an idealistic experimental data set (Fig. 3.4). Reporting biological parameters from a mathematical model should incorporate the uncertainty assessment regarding the initial conditions used, for example, in a sensitivity analysis with respect to initial conditions.

The scientific community can actively benefit from an advanced understanding of the sources of error. The purpose of spotting these sources is to avoid or at least to recognize them. Efforts in dealing with errors in parameter estimation shall be documented in publications with mathematical models to strengthen and support further development in the field. There are rooms for further refinement of estimation methodologies to minimize the risk of reporting erroneous estimates and different types of measurement errors in experimental studies. Above all, however, there is a need to unify the efforts in modeling practices, such as to develop good practice guidelines for reporting mathematical model assessment results, communicate the difficulties, and construct a common software library. Ultimately, a reassessment of the current approach in modeling experimental data is needed to shape the future research on a more solid path, maintaining trust in the scientific community of the diligent work in modeling biological systems.

Part 2

Modeling Host Infectious Diseases

Chapter 4

Modeling Influenza Virus Infection

CONTENTS

4.1 INFLUENZA INFECTION

Influenza A virus (**IAV**) infections remain major causes of hospitalization, health complications, and death [71,72]. Previous outbreaks such as the Asiatic flu (H2N8) (1889–1890), Spanish flu (H1N1) (1918–1920), Asian flu (H2N2) (1957–1958), Hong Kong flu (H3N2) (1968–1969), and more recently the swine flu (H1N1) (2009) have revealed a high morbidity and mortality rate [72]. Particularly, influenza-related diseases are more severe among high-risk groups such as toddlers (<2 years old), pregnant women, seniors (≥65 years old), and immunocompromised individuals [72]. Influenza infection is influenced by continuous antigenic drifts, antigenic shifts, and host factors. Thus, vaccination strategies have to be reformulated continuously [73]. Untangling the immune system responses to influenza infection is central for the establishment of improved prevention, control, and treatment strategies. To this end, mathematical modeling is a promising tool to quantify influenza infection.

Influenza represents the most pathogenic strain to human host [74]. Influenza, an RNA virus, has a high mutation rate infecting mainly epithelial cells of the upper respiratory tract. However, influenza can infect cells of the lower respiratory tract in more severe cases [75]. The hemagglutinin (HA) is a main protein in viral envelope and is the determinant of virus pathogenic. HA protein initiates the infection with both binding and fusion capacity [74].

Influenza acute symptoms can be developed within one to four days after infection. Infected individuals become infectious from the day be-

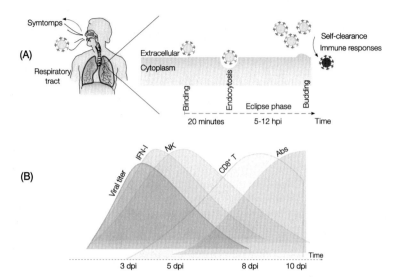

■ **FIGURE 4.1 Influenza A virus infection and immune system dynamics.** (**A**) Description of the main phases of influenza infection within a host. After entering the respiratory tract, each virion binds to a target cell. Then, virions enter the eclipse phase (5–12 hpi), before starting to replicate and infecting other cells; (**B**) Influenza virus and immune response dynamics. The innate response is mainly represented by interferon (IFN)-I and by natural killer (NK) cells, whereas the adaptive response is driven by cytotoxic CD8^{+} T cells (CTLs) and antibodies (Abs). Days post infection is abbreviated with dpi. Figure taken from [8].

fore the symptoms exhibit until seven days afterward. Typical symptoms of influenza include fever, muscle aches, fatigue, runny nose, and sore throat [76]. Influenza infections are considered self-limited due to the development of a protective immune response [77]. Fig. 4.1A shows major infection events within a host [78].

Influenza infection starts when virions enter the upper portion of the respiratory tract. In the first phase, the binding and fusion capacity of the hemagglutinin protein facilitates virion adhesion to the receptors of the epithelial cell surface (*binding*). Approximately 20 min after infection, the virion is endocytosed inside the epithelial cell (*endocytosis*) [74]. Once inside the cell, the viral RNA is released and used as a template to produce new virions, which are subsequently released into the extracellular environment (*budding*). The period between successful infection of the cell and the virus release is denominated the "*eclipse phase*", which ranges between 5 and 12 hours post infection (hpi) [7,79,80]. After that, the new virions can start to infect other cells. The loss of virions can be attributed either to loss of virus infectivity or clearance by the immune system.

Virus replication peaks around 2–3 days post infection (dpi). Viral clearance is observed in most of the cases between 5 and 10 dpi. Influenza specific antibodies (Abs) and cytotoxic $CD8^+$ T cells (CTLs) are detectable after 5 dpi and peak at 7 and 10 dpi, respectively [81]. However, Immunoglobulin (Ig)G Abs peaks later around 25 dpi [82]. Natural killer (NK) cells and interferon (IFN)-I are the main components of the innate immune system attributed to play a role in controlling influenza infections [44]. Fig. 4.1B depicts the dynamics of influenza infection and the principal elements of the immune system.

4.2 MODELING INFLUENZA INFECTION

Mathematical models have been developed with different grades of detail providing quantitative insights into viral dynamics within a host [7, 70]. Mathematical models have the potential to quantify the main immune factors involved in the responses to influenza infections dissecting critical information to advance understanding of the immunity to influenza. Importantly, mathematical models can suggest strategies to boost the immune response to influenza vaccines in populations that are prone to complications (e.g., the elderly). Two excellent reviews in influenza mathematical modeling are [7,8].

Table 4.1 presents the most recent list of works concerning mathematical models in influenza infections. Quantitative results can be largely affected by the host and the experimental setting. Animal models can allow more controlled experiments [74]. Influenza infection kinetics have been studied in different hosts, including avian species (wild bird and domestic poultry), mammals (swine, pony, ferret, mouse, guinea pig, nonhuman primates) [74]. The mouse model is widely used mainly because of its low cost, availability of different strains, and "knockouts" of genes that lead to deficiencies of specific immune populations or products [83].

In Vivo **setting.** The first mathematical model to describe influenza virus dynamics was developed in 1976 by [107]. The model was fitted to viral titer data of mice infected with influenza H3N2. After thirty years without modeling efforts, a pioneering work to describe influenza infection was presented by Baccam et al. [44] based on the well-known target cell model. The target cell model is represented by susceptible cells (U), infected cells (I), and virus (V) as shown in Fig. 4.2. Typically, the viral load peaks and subsequently declines once the uninfected cells are depleted.

$$\dot{U} = -\beta UV, \tag{4.1}$$

$$\dot{I} = \beta UV - \delta I, \tag{4.2}$$

$$\dot{V} = pI - cV. \tag{4.3}$$

Table 4.1 Mathematical models in influenza infection [8]

References	In Vitro	In Vivo Innate	In Vivo Adaptive	Host	Coinfection	Aging
Baccam et al. [44]		✓				
Beauchemin et al. [84]	✓					
Bocharov et al. [85]		✓				
Boianelli et al. [30]		✓				✓
Canini et al. [86]		✓				
Cao et al. [87]		✓				
Chen et al. [88]		✓				
Dobrovolny et al. [89]	✓			Various		
Duvigneau et al. [33]		✓	✓			✓
Hancioglu et al. [90]		✓				
Handel et al. [91]		✓	✓			
Handel et al. [92]			✓			
Heldt et al. [69]	✓					
Holder et al. [93]	✓					
Holder et al. [94]	✓					
Lee et al. [95]		✓	✓			
Miao et al. [82]		✓	✓			
Mohler et al. [96]	✓					
Paradis et al. [97]	✓					
Pawelek et al. [98]		✓				
Petrie et al. [99]		✓				
Pinilla et al. [80]	✓					
Price et al. [100]		✓	✓			
Reperant et al. [101]		✓	✓			
Saenz et al. [102]		✓				
Schulze et al. [103]	✓					
Smith et al. [104]		✓			✓	
Smith et al. [105]		✓			✓	
Tridane et al. [106]			✓			
Vargas et al. [50]		✓	✓			✓

Several mathematical works have tried to model the eclipse phase in vivo [44] and in vitro [84,94], representing the time frame of the infection more adequately. This has resulted in an additional state, in which newly infected cells rest in a latent phase before becoming productively infected cells (I). Thus, the model in Eqs. (4.1)–(4.3) with the eclipse phase can be represented as follows:

$$\dot{U} = -\beta U V, \qquad (4.4)$$

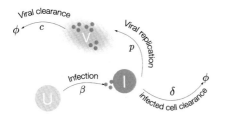

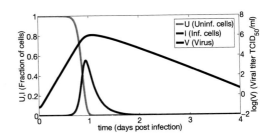

$$\dot{E}' = \beta U V - k E', \tag{4.5}$$

$$\dot{I} = k E' - \delta I, \tag{4.6}$$

$$\dot{V} = p I - c V, \tag{4.7}$$

where E' represents the cells in the eclipse phase, which can become pro-ductively infected at rate k. In [94], the authors considered different time distributions for modeling the eclipse phase and viral release by infected cells. The different mathematical model formulations were fitted with in vitro data. The results showed that the time distribution forms of the eclipse phase and viral release directly affect the parameter estimation.

The model equations (4.4)–(4.7) were fitted in [44] with data from human volunteers infected with influenza A/HongKong/123/77 (H1N1). The esti-mated biological parameters, for example, viral clearance and cell half-life, provided quantitative means of the viral infection dynamics. In a similar di-rection, Handel et al. [91] applied the target cell model to human influenza data to assess the emergence of resistance to neuraminidase inhibitors. How-ever, this study also raised identifiability issues, and confidence intervals were not provided. Recently, Dobrovolny et al. [108] adopted a double target cell model with the eclipse phase to predict the efficacy of neuraminidase in-hibitors on uncomplicated and more severe infections inside a host. For this purpose, the authors considered two different types of target cells (default and secondary). The first type is the fraction of cells available to influenza infection, whereas the second type of cells is the one accessible for severe influenza infections. The model was capable of mimicking the dynamics of uncomplicated influenza infection, including the immune response in the secondary cell population dynamics.

In another study [99], parameter uncertainty was investigated using the tar-get cell model and explored the possible reduction of parameter uncertainty fitting by measuring both infectious (via tissue infection culture dose 50 ($TCID_{50}$)) and total viral load (via reverse transcription polymerize chain reaction (RT-PCR)). In addition, the variation in $TCID_{50}$ assay sensitivity and calibration may affect the parameter estimation [99].

■ FIGURE 4.2 Target cell model for influenza infection. (Left) Influenza (V) infects susceptible cells (U) with rate β. Infected cells are cleared with rate δ. Once cells are productively infected (I), they release virus at rate p, and virus particles are cleared at rate c. The symbol ϕ represents clearance; **(Right)** Computational simulations of the target cell model. Parameter values used for model simulation are taken from [44]. The susceptible cells (red (mid gray in print version) line) are rapidly infected, whereas the virus (black line) and infected cells (blue (dark gray in print version) line) peak at day one approximately. The viral growth is limited by the number of susceptible cells, decreasing the viral load and the number of infected cells to undetectable levels [8].

Immune response to influenza. Although the target cell model can predict the dynamics of influenza infection without considering the host immune response, understanding host responses to the viral infection is fundamental [109]. To better understand the factors shaping the influenza infection course, a comprehensive model should incorporate the immune response dynamics and its interactions with influenza. Mathematical models were developed to investigate and predict the immune responses of the host during influenza infections [70].

The first model was proposed by [85], which uncovered that CTLs and Abs are the main players controlling influenza infections. Adopting the same modeling approach as [85], Hancioglu et al. [90] concluded that the initial viral load influenced disease severity. The target cell model was extended by Baccam et al. [44] to include the role of IFN-I. The IFN-I dynamics were modeled with the equation $\frac{dF}{dt} = sE'(t - \tau) - \alpha F$, where F is the level of the IFN-I, s is the secretion rate of IFN-I by infected cells, α is the IFN-I clearance rate, and τ is the lag period necessary to secret IFN-I. The influence of IFN-I was assumed to inhibit the viral replication rate through the fractional form $\frac{p}{1+\epsilon F}$, where p is the viral replication rate, and ϵ is the effectiveness of IFN-I. Although problems of parameter identifiability arose, the mathematical model including IFN-I dynamics was able to describe the double peaks present in some viral titer data.

The role of IFN-I in preventing the infection of new cells was also investigated in [50,87,90,98,102]. In these works, the target cell model was extended taking into account another compartment where the cells remain refractory to influenza infection. In [90], the dynamics of IFN-I were considered to promote the resistance of epithelial cells to influenza infection. An increased production rate of IFN-I and induction to resistance rate were fundamental to control disease duration and damage.

Saenz et al. [102] modeled viral and IFN-I dynamics using data from equine influenza infection, revealing the extensive role of innate immunity in controlling the rapid peak of influenza infection. Using the same equine influenza infection data but with fewer parameters than [102], Pawelek et al. [98] showed that the rapid and viral decline after the peak can be explained by the killing of infected cells mediated by IFN-I activated cells, such as NK cells. Vargas et al. [50] revealed that proinflammatory cytokines levels (IFN-α, IFN-γ, tumor necrosis factor (TNF)-α) could be responsible for slowing down viral growth in aged mice, limiting the activation of CTLs and causing the impaired immune response to influenza infection reported in elderly.

Host reinfection models including different IFN-I control mechanisms of viral infection on in vivo data were compared [87]. The authors revealed that a model that included a cell in an IFN-induced state-of-resistance cannot explain the observed viral hierarchy. Moreover, the authors postulated that the dynamics of secondary infection strongly depend upon the interexposure time. Another relevant component of the innate immune response is driven by NK cells. Canini et al. [86] introduced a mathematical model with NK activation induced by IFN-I. Activated NK cells (N) follow the equation $\frac{dN}{dt} = F - \rho N$, where ρ is the NK cell death rate, and F is the level of IFN-I. Fitting of the viral kinetic and symptom data showed a correlation between within-host parameters and illness course.

The adaptive immune response dynamics directed by CTLs and Abs have been investigated in different works, all of which have concluded that CTLs play a relevant role in the clearance of the viral infection [70,82,90,92,95, 100]. The mathematical model in [95] predicted that CD4$^+$ T cells have a prominent role in antibody persistence and CTLs in the lung are as effective as neutralizing Abs for virus clearance, when present at the time of challenge.

A relevant study dissecting the early and late influenza infection phases was conducted by Miao et al. [82], estimating important kinetic parameters, for example, the half-life of infected cells or the infection rate of target cells. Using a simplified Abs response in [110], the authors argued that the data considered were not sufficient to discriminate the effect of the immune responses. A modeling study by Dobrovolny et al. [70] assessed the ability of mathematical models including different components of the immune system to predict the viral titer course. The authors compared these models with "knockout" experimental data, finding that no single model was able to explain different viral kinetics. Recently, a within-host mathematical model was developed by Price et al. [100], merging the innate and adaptive immune dynamics with inflammation. Authors showed a positive effect of controlled inflammation by the immune response.

In Vitro **setting.** The complexity of the host immune system poses a significant challenge for the identification of critical factors involved in the interaction with influenza. This complexity has limited the exploration of some of their basic biological functions. An in vitro setting can simplify the system providing a small number of components to study. For example, key biological aspects can be inferred by analyzing viral infections of cell cultures. In fact, a large array of mathematical models, focused on understanding the interactions between the virus and target cells, have been developed using in vitro generated data. For instance, [96] applied the target cell model to include cell death and replenishment of target cells. This model described the viral kinetics of the equine influenza virus (A/Equi 2 (H3N8), Newmarket 1/93) in Madin–Darby canine kidney (MDCK) cells (microcarrier culture) and concluded that the number of available target cells is fundamental for viral growth. The dynamics of MDCK cell infection by influenza virus (A/Albany/1/98 (H3N2)) were studied in the presence of amantadine, which showed that the efficacy of this drug to block the infection of target cells was 56%–74% [84]. In [103], infection of MDCK cells with different influenza strains was studied. The model described the dynamics of virus particle release (infectious virions and hemagglutinin content), and the parameters were estimated for different virus strains. The results suggested that strains with slow initial infection dynamics but late induction of apoptosis resulted in higher viral yields.

In [93], the fitness of the wild-type (WT) influenza virus (A/Brisbane/59/2007 (H1N1)) was compared to the strain that acquired the oseltamivir-resistance mutation H275Y in its neuraminidase (NA) gene. Two different experimental settings were used, plaque assay and viral yields. The plaque assay indicated a higher fitness for the WT strain. In contrast, the viral yield assay suggested a higher fitness for the H275Y strain. Interestingly, in silico versions of these assays showed that the plaque assay was more sensitive for studying the duration of the eclipse phase, whereas the viral yield assay was sensitive for virus production. The different sensitivities of two assays could explain the conflicting results in terms of fitness. Later, results by Pinilla et al. [80] revealed that the principal effects of the H275Y substitution on the pandemic H1N1 (H1N1pdm09) strain were to lengthen in the mean eclipse phase of infected cells (from 6.6 to 9.1 h) and decrease (by 7-fold) the viral burst size.

More recently, using the same model, the viral fitness of H275Y and I223V NA mutants were compared with the pandemic H1N1 (H1N1pdm09) WT strain [97]. The eclipse phase duration was longer for both mutants than the eclipse phase length of WT strain (by 2.5 and 3.6 h, respectively). These works show the enormous benefits of mathematical modeling and how in vitro experimental data can support the development of a mathematical model and vice versa.

A major problem when comparing mathematical models is the difference between experimental settings in which the experiments are performed. For example, viral titers are provided in different units (e.g., RNA copies, plaque-forming units (PFU/mL), focus-forming units (FFU/mL), $TCID_{50}$, or egg infection dose 50 (EID_{50}/mL)). Even in the case where experimental data present the same viral titer units, experiments conducted in different labs cannot be compared because the viral units are not standardized [97].

Paradis et al. [97] suggested that even similar experiments performed in the same laboratory but at different time points cannot be compared. In addition to the noisy nature of biological measurement, the data used in mathematical models were not usually collected with a modeling purpose in mind. As a consequence, the data collected may not be optimal for identifying the model parameters. In particular, there are usually many measurements in the viral growth phase, whereas measurements in the clearance phase are missing. These issues could be solved if a close and iterative collaboration between biologists and modelers was established before the experimental design.

4.3 INFLUENZA INFECTION MODEL WITH IMMUNE RESPONSE

Several studies investigated quantitatively the relevance of CTLs to clear influenza infections [82,95,98,102]. However, due to the identifiability limitations to estimate the parameters in the target cell model only with viral load data, Boianelli et al. [8] proposed a minimalistic model able to fit influenza and CD8+ T dynamics (Fig. 4.3).

■ **FIGURE 4.3 Viral infection model with CTLs response.** Influenza (V) induces CTLs (E) clonal expansion with rate r, which inhibits the viral replication through the clearance of the infected cell; this effect can be included in c_v. CTLs are replenished with rate s_E and die with rate c_e. Figure taken from [8].

It is assumed that the clearance of the infected cells by CTLs can be represented by the clearance of the virus. The dynamics of V and E can be described by the following equations:

$$\dot{V} = pV\left(1 - \frac{V}{K_v}\right) - c_v V E, \tag{4.8}$$

$$\dot{E} = rE\left(\frac{V}{V + k_e}\right) - c_e E + s_E. \tag{4.9}$$

The CTLs replenishment rate was defined as $s_E = c_e E_0$, where E_0 is the initial number of CTLs, and c_e is the half-life of CTLs. The steady state should be satisfied to guarantee the CTLs homeostatic value $E = s_E/c_e$ in the absence of viral infection ($V = 0$). The CTLs proliferate at rate r. It is assumed the activation of CTLs proliferation by influenza follows a Michaelis–Menten growth with half saturation constant k_e. The virus growth is modeled with a logistic function with a maximum carrying capacity K_v and growth rate p. The virus is cleared at rate $c_v E$.

For parameter fitting, it was considered the murine data from [61] to estimate the model parameters. In this cross-sectional study, young and aged BALB/c mice were infected with 50 μL (50–100 PFU) of influenza H1N1(PR8). Lung viral titers were measured at different time points by plaque assays. Various components of the immune system were also monitored in [61]. To reduce the number of parameters to be identified, Boianelli et al. [8] fixed the value of $c_e = 2 \times 10^{-2}$ (half life is approximately $T = 34$ days) and K_v equal to the maximum value of the viral titer data.

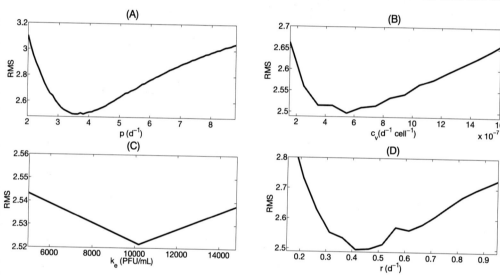

■ FIGURE 4.4 Profile likelihood for the model parameters. (A) p is the viral replication rate. **(B)** c_v represents the viral clearance. **(C)** k_e is the CTLs half saturation constant. **(D)** r represents the CTLs proliferation rate. Figure taken from [8].

Consequently, the profile likelihood analysis was computed for the unknown model parameters; this is shown in Fig. 4.4. Note that all the parameters present a profile likelihood with a minimum, implying that parameters are identifiable [63]. Therefore, the results suggest the possibility to infer model parameters from the available experimental data set.

A thousand sampling repetitions from the data in [61] were performed [8]. In each repetition, a random sample from each time point was created, and the parameters were estimated. Model fitting and parameter distributions from nonparametric bootstrap are shown in Figs. 4.5 and 4.6, respectively. Parameter estimates and 95% confidence intervals are shown in Table 4.2, respectively. The constraints for model parameter optimization, the lower and upper bounds, were broadly chosen to consider a large biological parameter space (see Table 4.2). The quality of fitting, as shown in Fig. 4.5,

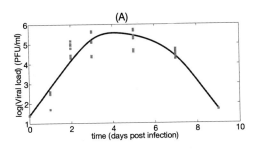

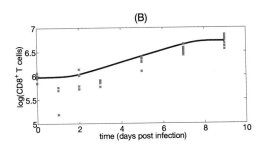

Table 4.2 Model parameter estimates (median)			
Parameter	**Median**	**Confidence Interval (95%)**	**Constraints for Optimization Algorithm**
p [d^{-1}]	4.4	[3.43; 6.08]	[1; 8]
c_v [d^{-1} cell^{-1}]	1.24×10^{-6}	[6.1×10^{-7}; 2.73×10^{-6}]	[5×10^{-8}; 10^{-5}]
r [d^{-1}]	0.33	[0.20; 0.42]	[0.01; 1]
k_e [PFU/ml]	2.7×10^3	[5.10×10^2; 1.06×10^4]	[4×10^2; 3×10^4]

Notes: The initial conditions of the model are $E_0 = 10^6$ cells, and $V_0 = 25$ PFU/mL.
$c_e = 2 \times 10^{-2}$ (half life is approximately 34 days).
$k_v = 10^6$ PFU/mL.

■ **FIGURE 4.5 Viral infection model fitting.**
(**A**) The model fitting is shown with a blue (dark gray in print version) line, and viral load data is presented in red (mid gray in print version) squares for mice infected with the H1N1 (PR8) strain. (**B**) the model fitting is shown in a blue (dark gray in print version) line, and the CTL data is presented in red (mid gray in print version) squares. Figure taken from [8].

indicates that the model is able to describe murine data. The computed confidence intervals in Table 4.2 indicate a large interval only for the parameter k_e, whereas they result in a low extent for the remaining parameters.

The large confidence interval of the parameter k_e may be attributed to the flat profile likelihood observed in Fig. 4.4C. Quantitative comparisons of parameters in Table 4.2 with the target cell model parameters are not possible due to the differences in model structures and units. Smith et al. [111] proposed an approximation for the viral growth, which is equivalent to Eq. (4.8) in the growth phase. The p estimates [3.43; 6.08] d^{-1} are in the same range of the estimate (6.59 d^{-1}) in [111]. Moreover the range of viral clearance rate $c_v E$ [1; 5.27] d^{-1} is consistent with previous estimates [2.6; 15] d^{-1} [44]. To explore possible parameter dependencies, scatter plots were built from the computed parameters values using the nonparametric bootstrap as it is shown in Fig. 4.7. The scatter plots in Fig. 4.7 show interdependencies between p-c_v, p-r, and c_v-r parameters. This implies that the estimation of one parameter influences the estimate of the others. For example, in the case of the parameters p-c_v, increasing the value of p increases the value of c_v, whereas increasing the value of p decreases the value of r. These parameter correlations could lead to possible parameter estimation problems. Moreover, the dependence of model parameter estimates p, c_v, r, k_e can be evaluated to viral carrying capacity K_v. The estimates of p and r showed

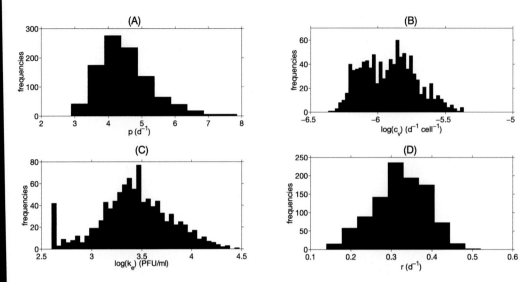

■ FIGURE 4.6 Nonparametric bootstrap results. The distributions of the nonparametric bootstrap obtained with 1000 samples for the model parameters (**A**) p, (**B**) c_v, (**C**) k_e, (**D**) r. Figure taken from [8].

no correlation with the parameter K_v, whereas there is a positive correlation of c_v and k_e estimates with K_v. Ad hoc experiments to estimate parameter values could be a possible solution to alleviate parameter interdependencies and, therefore, estimate the remaining parameters more adequately. For example, in the previous work proposed by [80], virus titers were incubated without target cells and followed up to determine the remaining infectious titers. In this way, the approximate values of the viral clearance rate could be determined providing more accurate estimates.

4.4 POSTINFLUENZA SUSCEPTIBILITY TO PNEUMOCOCCAL COINFECTION

Retrospective studies performed on victims of the 1918/1919 influenza A virus pandemic and also the recent H1N1 influenza pandemic revealed a high incidence of coinfections with unrelated bacterial pathogens [112, 113]. In fact, 71% of the high death toll during the 1918/1919 outbreak was attributed to coinfection with *Streptococcus pneumoniae* (*S. pneumoniae*) [113]. This copathogen is a human adapted gram-positive colonizer of the nasopharynx in asymptomatic children, adults, and individuals over 65 years [114]. Animal and human studies have shown that preceding influenza infection enhances all aspects of *S. pneumoniae* pathogenesis from nasopharyngeal colonization to invasive pneumococcal disease [115], leading to the strong predisposition to lethal secondary pneumococcal infection in influenza infected patients. The underlying mechanisms implicated in this synergism include physical disruption of the lung epithelial barrier and aberrant immunological responses causing lung injury.

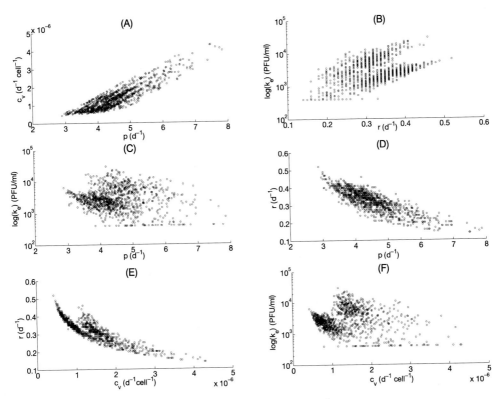

Altogether coinfection studies strongly reflect detrimental changes in lung microenvironment primary or secondary to the exacerbated cytokines production. Nevertheless, it is still largely unknown whether these cytokines work alone or in synergism as "friend or foe" to the coinfected host [116]. In the absence of pieces of the puzzle, mathematical modeling can serve as a framework to test different hypotheses and tailor future experimental settings. Smith et al. [104] proposed the first modeling approach to represent the basic interactions between IAV and bacterial coinfection by *S. pneumoniae*. Experimental and theoretical work to dissect the contributions of proinflammatory cytokines as main responsible candidates for the colonization in a secondary bacterial infection was presented by [33].

Mouse infection experiments. Mouse infection experiments are taken from previous work in [33]. Female C57Bl/6J wildtype mice were purchased at the age of 8 to 9 weeks. The influenza A virus strain A\PR8\34 (H1N1) [117] and the encapsulated *Streptococcus pneumoniae* strain TIGR4 (T4) were employed.

The measurements at the time points 0 h, 6 h, 18 h, 26 h, and 31 h were reported in three independent experiments. For every time point, the follow-

■ **FIGURE 4.7 Scatter plot results.** The scatter plots of (**A**) p-c_v, (**B**) k_e-r, (**C**) k_e-p, (**D**) r-p, (**E**) r-c_v, and (**F**) k_e-c_v. The numerical values are obtained from the nonparametric bootstrap distributions in Fig. 4.6. The plots show dependencies between parameters p-c_v, p-r, and c_v-r. Figure taken from [8].

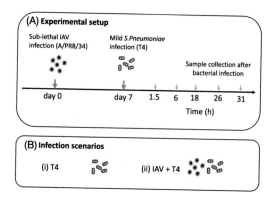

■ **FIGURE 4.8 Experimental setting.** C57Bl/6J wildtype mice were nasal infected with sublethal dose of IAV (A/PR8/34) following bacterial infection with the T4 strain of *S. pneumoniae* at day 7. Blood, bronchoalveolar lavages, and lung homogenates samples were taken from all animals at 1.5, 6, 26, and 31 hours (**A**). The infection scenarios are single bacterial infection (T4) and coinfection (IAV+T4) (**B**).

ing infection scenarios are considered: single bacterial infection (T4) and coinfection (IAV+T4) starting at day 7 post influenza A virus (IAV) infection. Every group consisted of 7 mice. A schematic representation of the experiments can be seen in Fig. 4.8. Before infection, day 0, all mice were narcotized with a ketamine solution. The IAV+T4 group was intranasally infected with 25 µL of a diluted virus suspension. The T4 and IAV+T4 groups were administered with 25 µL PBS (Phosphate Buffered Saline). Body weight and health status were monitored during the next 7 days. Mice that lost more than 25% of body weight were sacrificed [33].

Bacterial Model. Here, we consider the model of *S. pneumoniae* infection (B) proposed in [33] and described by the following equation:

$$\frac{dB}{dt} = rB\left(1 - \frac{B}{K_B}\right) - c_b B, \tag{4.10}$$

where r is the bacterial proliferation rate with maximum carrying capacity K_B. The growth rate ($r = 1.13\ \mathrm{h}^{-1}$) and the carrying capacity ($K_B = 2.3 \times 10^8$ CFU/ml) correspond to single *S. pneumoniae* infection from previous work [104]. Phagocytosis of the bacteria is considered by $c_b = 1.28$ for single T4 infection and $c_b = 0.72$ for the coinfected group as presented by [33].

Using similar reasoning as [104], it was assumed that only a proportion of the bacterial inoculum may reach the lung since some bacteria can be removed by mucocilliary mechanisms. Thus, 1000 CFU/mL *S. pneumoniae* was considered as inoculum to fit the model parameters in Eq. (4.10). Fig. 4.9 reveals that 18 h after second infection (hpsi) the bacterial burden

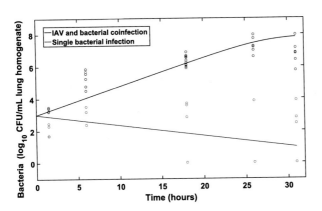

■ **FIGURE 4.9 Simulation results for Single bacterial infection and Coinfection.** Empty circles represent the experimental data taken from [33]. Coinfection takes place 7 days after influenza infection.

of coinfected mice is elevated in comparison to a single bacteria infected mice. Bacterial burden of T4 infected mice is not existent, although some of the single infected mice have slightly elevated CFU counts in BAL and lung samples. This dichotomy effect was also remarked previously by [104], where it was attributed to the host heterogeneity as it exists in humans. The peak for the CFU-counts of T4 infected mice is at 18 hpsi. Following this time point, the mean value for the bacterial burden of the T4 infected group decreases, whereas the measured values for coinfected mice are increasing. At later time points (after 18 hpsi), almost all mice of the coinfected group developed very high bacterial titer in the blood, which is an indicator for septicemia. Results in Fig. 4.9 highlight that many mice cleared the infection in the group infected only with bacteria.

Steady state and stability analysis. To this end, the bacteria equilibrium points (B^*) were derived by setting the right term of Eq. (4.10) equal zero:

$$0 = rB\left(1 - \frac{B}{K_B}\right) - c_b B. \qquad (4.11)$$

The first equilibrium ($B^{(1)}$) represented that there was no colonization of the bacteria:

$$B^{(1)} = 0. \qquad (4.12)$$

The second equilibrium point ($B^{(2)}$) was written as follows:

$$B^{(2)} = K_B\left(1 - \frac{c_b}{r}\right). \qquad (4.13)$$

The second equilibrium is biologically meaningful (positive definite) if $r > c_b$. This means that the bacterial proliferation rate needs to be higher than its clearance. Although this condition can be even followed by intuition, the hypothesis that influenza increases substrates as nutrients for bacterial growth (K_b) affects only the steady state that the bacteria will reach, but more nutrients do not contribute directly to the bacterial decision to colonize as can be observed in (4.13).

To evaluate if the increase in substrates modulated by influenza infection could alternatively alter the stability properties of system (4.10), the stability analysis was derived from the Jacobian of Eq. (4.10), which reads as follows:

$$\frac{\partial f_1}{\partial B} = (r - c_b) - 2\frac{r}{K_b}B^*. \tag{4.14}$$

For the first equilibrium point, the Jacobian is

$$\frac{\partial f_1}{\partial B}^{(1)} = (r - c_b). \tag{4.15}$$

Thus the equilibrium point $B^{(1)}$ will be stable if $c_b > r$. This implies that the bacteria will not colonize the host since system (4.10) has only one equilibrium point, which is zero and is attracted to it independently of the initial inoculation.

If $r > c_b$, then the equilibrium point $B^{(1)}$ is unstable, meaning that bacteria will grow independently of inoculation. Furthermore, the equilibrium point $B^{(2)}$ is biologically significant (bacterial colonization). By doing some algebra the Jacobian can be written as

$$\frac{\partial f_1}{\partial B}^{(2)} = (c_b - r). \tag{4.16}$$

From (4.16) it follows that the equilibrium point $B^{(2)}$ is biologically meaningful and attracting if $r > c_b$.

Nutrient availability modulation. Experimental evidence by [118] established that higher rates of disease during coinfection could stem from increased sialic acid availability, which further supports bacterial colonization and proliferation. From a mathematical point of view, nutrient sources can be represented by the current capacity K_B in Eq. (4.10). By the previous steady state and stability analysis, which is independent of parameter fitting procedures, it can be inferred that the bacterial nutrient source K_B determines the size of the initial bacterial colony but not the decision to grow. Of note, this approach is also valid assuming more complex logistic growth terms. For instance, the mathematical term $rB\left(1 - \frac{B}{K_B(1+\Psi V)}\right)$ previously proposed by [104] to integrate the increase in the carrying capacity $K_B(1 + \Psi V)$ may provide equivalent conclusions.

Role of proinflammatory cytokines in bacterial outgrowth. To dissect the temporal contribution of the measured proinflammatory cytokines in preventing bacterial clearance, mathematical models fitted from single *S. pneumoniae* infection in [33] were challenged to assess the effects of proinflammatory responses on bacterial lung titers. Hence, [33] adopted the best model for single *S. pneumoniae* infection and challenged the mathematical term representing bacterial clearance ($c_b B$) with different functions ($c_b f_x B$) to evaluate which of the proinflammatory cytokines or their combinations provided the best fit to the bacterial burden detected in the coinfected mice. Therefore, Eq. (4.10) was modified as follows:

$$\frac{dB}{dt} = rB\left(1 - \frac{B}{K_B}\right) - c_b f_x B,$$

where the phagocytosis of the bacteria is considered by the multiplicative term $c_b f_x$, where c_b is the constant phagocytosis rate, and the term f_x is the mathematical function that served to test different hypotheses. The experimental data for the different cytokines to build piecewise linear functions to dynamically determine the function f_x. A brief list of mathematical models tested in [33] is presented in Table 4.3. Considering the AIC scores, the criterion of small differences (less than 2 units) are considered not significant; the best group of models was M3 and M7 (Table 4.3). The common component of these two models is the IFN-γ kinetics. Remarkably, a mechanism only based on the IFN-γ response (M3) provided a better fit than mechanisms based on only TNF-α (M4) or IL-6 (M5) even though a more conservative AIC criterion is considered (e.g., ≤ 10). In agreement with previous work [119], the model selection process dissected IFN-γ as a key and sufficient modulator in the impairment of bacterial clearance.

Combining tailored experimental data and mathematical modeling, Duvigneau et al. [33] suggested a strong detrimental effect of IFN-γ alone and in synergism with IL-6 but no conclusive pathogenic effect of IL-6 and TNF-α alone. These findings correlate well with Vargas et al. [50], suggesting that the increased levels of pro-inflammatory cytokines (the "inflammaging" state), in particular, IFN-γ, contribute to the reported impaired responses in people over 65 years of age. Thereby, IFN-γ plays a pivotal role in driving severe disease during primary influenza infection in the elderly and bacterial outgrowth during coinfections.

4.5 CONCLUDING REMARKS

Mathematical models have incorporated different levels of complexity trying to explain the fundamental aspects of influenza infection and integrating

Table 4.3 Coinfection models [33]

No	Hypothesis	f_x	Parameters	RSS	AICc
M1	Alveolar macrophages (AMs) dynamics $M_A(t)$ are sufficient to facilitate bacterial outgrowth.	$M_A(t)$	c_B	47.74	12.27
M2	Single change in bacterial clearance rate is enough to explain bacterial outgrowth.	M_A^*	c_B	21.84	-15.87
M3	IFN-γ responses alone can impair bacterial clearance facilitating bacterial colonization.	$\frac{A_{\text{IFN-}\gamma}}{\text{IFN-}\gamma(t)+A_{\text{IFN-}\gamma}}$	$A_{\text{IFN-}\gamma}$	16.13	$\mathbf{-26.78}$
M4	TNF-α responses alone can impair bacterial clearance facilitating bacterial colonization.	$\frac{A_{\text{TNF-}\alpha}}{\text{TNF-}\alpha(t)+A_{\text{TNF-}\alpha}}$	$A_{\text{TNF-}\alpha}$	39.82	5.74
M5	IL-6 responses alone can impair bacterial clearance facilitating bacterial colonization.	$\frac{A_{\text{IL-6}}}{\text{IL-6}(t)+A_{\text{IL-6}}}$	$A_{\text{IL-6}}$	27.22	-7.94
M6	A synergistic effect of the IFN-γ (X_1), IL-6 (X_2), and TNF-α (X_3) cytokine responses in facilitation of bacterial outgrowth.	$\prod_{i=1}^{3}\left(\frac{A_i}{X_i(t)+A_i}\right)$	$A_{\text{IFN-}\gamma}$, $A_{\text{IL-6}}$ $A_{\text{TNF-}\alpha}$	14.57	-25.82
M7	A synergistic effect of the IFN-γ (X_1) and IL-6 (X_2) cytokine responses in facilitation of bacterial outgrowth.	$\prod_{i=1}^{2}\left(\frac{A_i}{X_i(t)+A_i}\right)$	$A_{\text{IFN-}\gamma}$, $A_{\text{IL-6}}$	14.57	$\mathbf{-28.21}$

a Best models based on AICc difference lower than 2 units are in bold.

most of the available biological knowledge. These include (*i*) bacterial coinfections, (*ii*) aging of the immune system and the role in influenza infection, (*iii*) challenges for influenza vaccination, and (*iv*) host and influenza genetic factors (Fig. 4.10).

Immunosenescence is a complex biological process and therefore not the result of isolated events. Age-related changes and their effects on influenza virus infection dynamics were examined in [50]. Vargas et al. [50] considered different immune system components in young and aged mice:

■ **FIGURE 4.10 Influenza Challenges.** Influenza infection is controlled by the host immune response, which in turn is shaped by host genetic factors, previous infections, vaccination, and aging. Figure taken from [8].

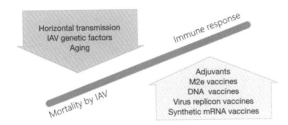

■ **FIGURE 4.11 Emerging vaccination strategies.** Novel vaccination strategies. For example, Novel technologies will enable rapid production of emerging virus strains (e.g., synthetic mRNA or RNA replicon-based vaccines) or universal vaccines covering major clades (e.g., designed hemagglutinins triggering broad neutralizing antibodies or vaccines triggering cross-protective CTL responses). Figure taken from [8].

CTLs, NK cells, IFN-I, IFN-γ, and TNF-α. The fits of mathematical models to the experimental data from young and aged mice suggested that the increased levels of IFN-I, IFN-γ, and TNF-α (the "inflammaging" state) could promote the slower viral growth observed in aged mice, which consequently could limit the stimulation of immune cells and contributes to the reported impaired responses in the elderly. This study illustrated the advantages gained by the mathematical modeling approach to dissect quantitatively the role of the different immune mediators in young versus aged. However, more efforts are needed to uncover the age-related changes. It is important to consider that preexisting immunity to influenza due to past infections/vaccination can directly alter the outcome described in [50].

Vaccination remains the cornerstone of influenza prophylaxis; however, the protection provided by seasonal vaccines is only partial, especially in high-risk groups such as the elderly [120]. Furthermore, new influenza virus strains arising from this zoonotic pathogen increase concerns about the threat of a new pandemic [121]. The global vaccine manufacturing capacity can be inadequate and unable to supply enough vaccine doses promptly [122]. Thus, to increase vaccination efficacy, huge efforts need to be taken to optimize vaccines that are already in the market and/or to develop new vaccination strategies (Fig. 4.11). Until now, mathematical models have focused mainly on the effect of the vaccines on population-level spread of influenza. However, quantitative efforts are necessary to gain comprehension into the immunological signatures of vaccination, evaluating the efficacy of different vaccines, adjuvants, delivery systems, and immunization routes.

Chapter

5

Modeling Ebola Virus Infection

CONTENTS

5.1 **EBOLA INFECTION**

Ebola was characterized for the first time in 1976 close to the Ebola River located in the Democratic Republic of the Congo [123]. Since then, outbreaks of **EBOV** among humans have appeared sporadically causing lethal diseases in several African countries, mainly in Gabon, South Sudan, Ivory Coast, Uganda, and South Africa [124]. Among the most severe symptoms of the EBOV disease are fever, muscle pain, diarrhea, vomiting, abdominal pain, and unexplained hemorrhagic fever [125]. Fatalities are predominantly associated with uncontrolled viremia and lack of an effective immune response. However, the pathogenesis of the disease is still poorly understood [126]. Ebola virus belongs to the family of *Filoviridae*, from Latin *filum*, which means thread [16]. Ebola virus is classified in Tai Forest, Sudan, Zaire, Reston, and Bundibugyo. The human Ebola epidemics have been mainly related to infection by the Zaire and Sudan strains. Filovirus virions possess several shapes (Fig. 5.1), a property called pleomorphism [126].

EBOV natural hosts still remain unsettled, but it is tenable that EBOV persists in animals, which transmit the virus to nonhuman primates and humans. It has been reported that fruit bats are capable of supporting EBOV replication without becoming ill and may serve as a major reservoir [127, 128]. EBOV can spread from an infected person to others through direct contact with blood or body fluids (e.g., saliva, sweat, feces, breast milk, and semen), objects (i.e., needles) contaminated with the virus and infected fruit bats or primates [126].

Modeling and Control of Infectious Diseases in the Host. https://doi.org/10.1016/B978-0-12-813052-0.00016-6

85

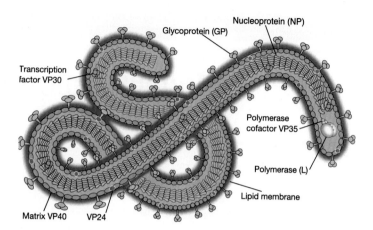

■ FIGURE 5.1 Ebola virus structure. The Ebola genome is composed of 3 leader, nucleoprotein (NP), virion protein 35 (VP35), VP40, glycoprotein (GP), VP30, VP24, polymerase (L) protein and 5′ trailer. Figure taken from [45].

The 2014 Ebola epidemic is the largest ever reported in history, affecting multiple countries in West Africa and imported to other countries [129]. Ebola virus can infect a wide variety of cell types including monocytes, dendritic cells, macrophages, endothelial cells, fibroblasts, hepatocytes, adrenal cortical cells, and several types of epithelial cells, all supporting EBOV replication. Monocytes, macrophages, and dendritic cells are early and preferred replication sites of the virus [126]. Furthermore, murine studies [130] have revealed that EBOV can infect cells in different compartments, showing high viral titers in liver, spleen, kidney, and serum. Animal models are pivotal to shed light on this lethal disease. Due to very close similarities with the human immune system, nonhuman primates are the preferred animal model for several viral infections such as HIV. Moreover, EBOV infection has been adapted to guinea pigs and mice [126], serving as a flexible model in comparison to human and nonhuman primates.

Due to its high infectivity and fatality, the virus is classified as a biosafety level-4 agent, restricting basic research for Ebola disease [131]. Therefore infection parameters and quantification of the interactions between the virus and its target cells remain largely unknown. Thus mathematical models are relevant tools to advance knowledge of EBOV infection. The first within-host mathematical model of EBOV infection was proposed by Nguyen et al. [45], who considered EBOV kinetics in a Vero cell line [131]. This chapter presents the most recent mathematical modeling work in Ebola virus infection, which is taken from [45,57].

5.2 **IN VITRO EBOLA VIRUS INFECTION MODEL**

The mathematical model proposed by Nguyen et al. [45] was based on the well-established target-cell-limited model; see details in Box 2.8. The EBOV infection model is as follows:

$$\frac{dU}{dt} = \lambda - \rho U - \beta U V, \tag{5.1}$$

$$\frac{dI}{dt} = \beta U V - \delta I, \tag{5.2}$$

$$\frac{dV}{dt} = p I - c V. \tag{5.3}$$

EBOV target cells can be in either a susceptible (U) or an infected state (I). Cells are replenished with constant rate λ and die with rate ρ. Note that the condition $\lambda = U_0 \rho$ should be satisfied to guarantee homeostasis in the absence of viral infection, so that only ρ is a parameter to be determined. Virus (V) infects susceptible cells with rate β. Infected cells are cleared with rate δ. Once cells are productively infected, they release virus at rate p, and virus particles are cleared with rate c. The initial number of susceptible cells (U_0) can be taken from the experiment in [131] as 5×10^5. The initial value for infected cells (I_0) is set to zero. The viral titer in [131] is measured in foci-forming units per milliliter (ffu/ml). The initial viral load (V_0) is estimated from the data using the fractional polynomial model of second order [132]. The parameter ρ is fixed from the literature as 0.001 day^{-1} [96]. The effect of fixing this value on the model output is evaluated with a sensitivity analysis.

Experimental data. Replication kinetics of EBOV can be studied in Vero cells, a cell line derived from kidney epithelial cells of African green monkeys [131]. This nonhuman primate is a known source of *Filoviridae* virus infection, for example, the European Marburg outbreak from 1967. Wild-type Vero cells and a Vero cell line expressing VP30 were tested to reveal their ability to confine EBOV to its complete replication cycle. Viral kinetics of wild-type Vero cells infected with EBOV at different multiplicities of infection (MOI) were considered [131]. The viral growth data is presented in Fig. 5.2.

All parameters in model (5.1)–(5.3) were shown to be algebraically identifiable given measurements of viral load and initial conditions [51]. However, the difference between structural identifiability and practical identifiability in the presence of measurement error requires further identifiability studies. To address practical identifiability, the approach proposed by [63] was considered in [45] for the data presented in Fig. 5.2. The resulting RMS profiles

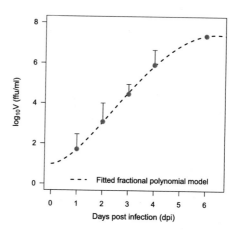

in Fig. 5.3 for β, p, and c show a convex shape of which the optimization routine can reach their minimum. Note that the profile of δ is flat in one tail, suggesting that the parameter δ can be chosen arbitrarily small without affecting the fit quality [63]. In spite of this, the lower bound of this parameter has a clear biological constraint. To be precise, the half-life of an infected cell cannot be longer than that of an uninfected cell. There is experimental evidence that the half-life of epithelium cells in the lung is 17–18 months in average [133]. In view of this, the infected cell death rate (δ) is fixed at 10^{-3}.

Bootstrapping can provide more insights into the distribution of parameter values based on experimental data in [131]. For clarity, only the weighted bootstrap [134] was presented in [45]. The distributions of the model parameters are shown in Fig. 5.4. Bootstrap estimates for the viral clearance (median $c = 1.05$ day^{-1}) are slightly below than other viral infections (Fig. 5.4). For example, clearance of influenza virus varied from 2.6 to 15 day^{-1} in [44, 50,98]. This may be attributed to the fact that the viral clearance is computed from in vitro experiments. EBOV is known to replicate at an unusually high rate that overwhelms the protein synthesis of infected cells [131]. Consistent with this observation, bootstrap estimates revealed a very high rate of viral replication, $p = 62$ (95% CI : $31 - 580$) (Fig. 5.4). Although the scatter plot in Fig. 5.4 shows that the estimate of p can be decreased given a higher effective infection rate β, a replication rate of at least 31.8 ffu/ml cell^{-1} day^{-1} is still needed to achieve a good fit of the viral replication kinetics in Fig. 5.2.

Scatter plots are a graphical sensitivity analysis method, and a simple but useful tool to test the robustness of the results. The estimated parameters are

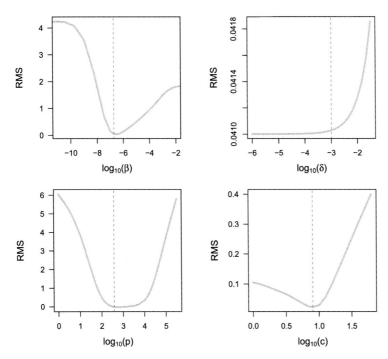

■ **FIGURE 5.3 Parameter Identifiability.** RMS profile of model parameters. Each parameter is varied in a wide range around the optimized value. Subsequently, the DE algorithm is used to refit the remaining parameters to the data set of [131]. The vertical dashed lines indicate the value obtained from the optimization for all four parameters collectively. Figure is taken from [45].

Table 5.1 Estimates of infection parameters [45]

Parameters (units)	Best fit	Bootstrap estimates		
		2.5% quantile	Median	97.5% quantile
β (day^{-1}ffu/ml^{-1})[10^{-7}]	1.91	1.78	4.06	261.95
p (ffu/ml day^{-1} cell^{-1})	378	31.80	62.91	580.69
c (day^{-1})	8.02	0.18	1.05	18.76
t_{inf} (hours)	5.64	1.68	9.49	10.79

plotted against each other. Scatter plots for the parameters in Fig. 5.4 provide visual evidence that these parameters strongly depend on one another, so that their individual values cannot be independently determined. That is, increasing the values of p increases the estimations of c, and decreasing the estimations of β increases the estimation of both c and p. However, the green (dark gray in print version) curves in Fig. 5.4 provide the most likely region where the parameters values can be found. To verify this intuition, in [45] the viral clearance rate c was fixed at 4.2, and then the other two

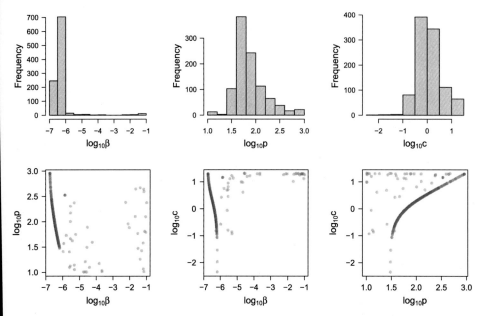

■ **FIGURE 5.4 Weighted bootstrap results.** *Top row*: Distributions from 1000 sample estimates are presented for the three parameters: β, p, and c. *Bottom row*: Scatter plot between bootstrap parameters. The parameter ρ is fixed during the bootstrapping at 0.001 [96]. Numerical values for model (5.1)–(5.3) are presented in Table 5.1. Figure is taken from [45].

parameters β and p were estimated. The results of 1000 bootstrap replicates reveal that fixing the parameter c improves the fitting with a narrow confidence interval.

In model fitting, some parameters may have little effect on the model outcome, whereas other parameters are so closely related that simultaneous fitting could be a difficult task. For this aspect, the scatter plots using pairs of parameters over different bootstrap replicates were reported in [45]. Sensitivity analysis was also performed [135]. For each data point, the derivative of the corresponding modeled variable value with respect to the selected parameter was computed. The normalized sensitivity function reads as

$$\frac{\partial y_i}{\partial \Theta_j} \cdot \frac{w_{\Theta_j}}{w_{y_i}}, \tag{5.4}$$

where y_i denotes the model variables, Θ_j is the parameter of interest, and the ratio w_{Θ_j}/w_{y_i} is the normalized factor corresponding to its nominal value [136]. Summary statistics of the sensitivity functions can be used to qualify the impact of the parameter on the output variables, that is, the higher the absolute value of the sensitivity summary statistics, the more important the parameter [135]. The sensitivity functions are plotted versus time to illustrate the role of parameters for the model output. The parameters that have little effect do not need to be fine-tuned extensively in model fitting.

Figs. 5.5A–E show the effect on the viral load when varying the respective parameter by 10%, 20%, and 50% around its nominal value. It can be seen

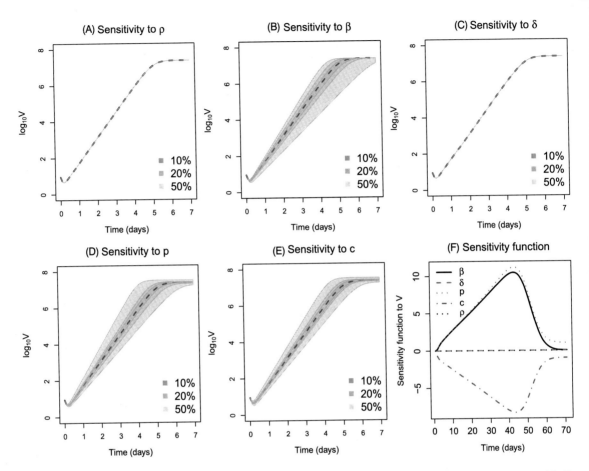

■ **FIGURE 5.5 Parameters sensitivity.** *(A)–(E)*: Plotting of viral titer variation versus time. The dashed line is the viral kinetics obtained from nominal parameter values. Three color shades in each figure represent the viral load variation range when varying the corresponding parameter by a percentage denoted in the legend. *(F)*: Parameter sensitivity function over time, the values in y-axis are calculated using Fig. 5.4. Figure is taken from [45].

that the host cell death rate ρ, which in the virus-free steady state represents the cell turnover, has little effect on the viral load kinetics. This can be attributed to the fact that the experiment was performed in vitro and within a short period. Similarly, the effect of the infected cell death rate δ can also be neglected. This could be explained by the fact that the observed Ebola viral load was not decreasing (Fig. 5.2), contrary to observations in other viral infections, for example, influenza virus [44]. The remaining three parameters β, p, and c are sensitive in the sense that a small change in parameter value can lead to a large difference in viral kinetics. Fig. 5.5F summarizes in detail the parameter sensitivity functions. The parameters β, p, and c govern the infection kinetics, whereas the effect of the parameters ρ and δ can be neglected for this data set.

Moreover, both β and p can be seen as consistently increasing the viral load because their respective sensitivity functions are always positive, in contrast to the parameter c. Note that the absolute magnitude of change in the sensitivity functions of these three parameters is approximately equal over time (Fig. 5.5F). The strong similarity in the sensitivity functions indicates that the corresponding parameters have an equivalent effect on the viral titer. For instance, the sensitivity functions of β and p are very similar so that almost the same output of viral titer will be generated by increasing β if p is decreased correspondingly. A similar statement can also be made about the relationship between c and β.

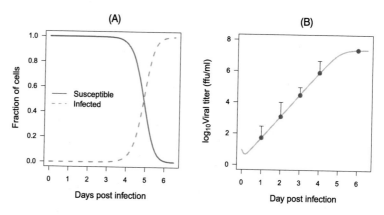

■ **FIGURE 5.6 Model fitting for EBOV kinetics.** Viral titer data with low MOI from [131] and simulations from the best fit shown in Table 5.1 are in panel (A) for the host cells and (B) for the viral titer. Figure is taken from [45].

Computational simulations for the best fitting is plotted in Fig. 5.6B, which shows that the virus grows exponentially from day 1 to 5 post infection. This is consistent with the mathematical analysis developed in [137], which deduced that the virus initially grows exponentially and can be better modeled as $\exp(r_0 t)$, whereas the susceptible cell population remains relatively constant, where r_0 is the leading eigenvalue that solves the equation $r_0^2 + (c + \delta)r_0 - (\beta p U_0 - c\delta) = 0$. Viral titer peaks at high levels, more than 10^7 ffu/ml, which in general is 10 fold higher than those reported in influenza virus infection [61,50]. In addition, the virus titer reaches a plateau at day 6 and may remain at those levels (Fig. 5.6B). No depletion of infected cells is observed in the period of observation. This could be a combined effect attributed to either high infection rate or high replication rate, and to the slow clearance of infected cells. To achieve virus titer levels as reported in [131], either a high infection rate β of susceptible cells or a high replication rate is required (Fig. 5.4). Note that even though these estimations were performed in vitro, in vivo murine studies for EBOV infection [130] showed similar kinetics and time scales as those presented in Fig. 5.6B.

Transmission measures. Infectivity is a critical parameter to assess the ability of a pathogen to establish an infection [138]. To determine infectivity, Nguyen et al. [45] computed the reproductive number R_0, see details in Definition 2.33. This epidemiological concept can be applied to model (5.1)–(5.3) and computed as follows [45]:

$$R_0 = \frac{\lambda p \beta}{c \rho \delta}. \tag{5.5}$$

The estimated reproductive number in EBOV infection is very high; see Fig. 5.7A and numerical results in Table 5.1. These results can be attributed to the fact that no depletion of virus was observed and to a slow clearance of infected cells. Thus, both parameters δ and c increase the value of R_0. Note that very high estimates of the reproductive number in highly viremic influenza virus strains from in vitro experiments have also been reported, with an average of 13×10^3 [80]. It is worth mentioning that fitting the model to in vitro data in [131] could lead to small estimates for c and δ in comparison to an in vivo situation. Nevertheless, estimates of the epithelial cell half-life were 6 months in the trachea and 17 months in the lungs in average [133], which corresponds to δ equal to 0.003 and 0.001, respectively. As mentioned previously, δ was fixed at 0.001 in the computation of R_0. Therefore, the estimated values of R_0 interval are very likely to be positioned in a biologically plausible range, especially, the upper bound. Notwithstanding, the estimate of R_0 presented in [45] should be interpreted with care within the limits of the data used.

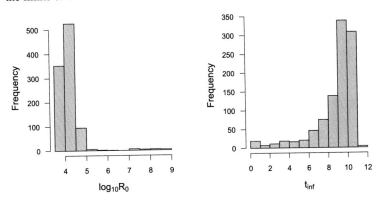

■ **FIGURE 5.7 Transmission measures.** Bootstrap estimate of the reproductive number (R_0) and the infecting time (t_{inf}) in hours. Numerical values can be found in Table 5.1. Figure is taken from [45].

Recent viral modeling works [80,94] have also introduced the term *infecting time*, which represents the amount of time required for a single infectious cell to cause the infection of one more cell within a completely susceptible population. Strains with shorter infecting time have a higher infectivity [80, 94]. This measure can be computed as $t_{\text{inf}} = \sqrt{\frac{2}{p\beta U_0}}$.

Bootstrap results showed that EBOV possesses an average infecting time of 9.49 hours (Fig. 5.4), which is approximately 7 times slower than the infecting time of influenza virus [94]. This number provides a reasonable explanation for the kinetics of susceptible cells, which slowly decrease from day 1 to day 4 (Fig. 5.6A) and quickly deplete within the last two days. This number can also explain the absence of viral replication within the first 5.6 hours after infection. This period corresponds to the short decreasing period observed in Fig. 5.6B. The initial decrease of viral load thus can be attributed to self-clearance of the virus when some viruses have infected cells but are not yet able to replicate.

The infectivity parameters in Fig. 5.7 characterize the EBOV infection kinetics in the data in [131]. The slow infection time of EBOV is compensated by its efficient replication. As a result, a short delay is followed by a massive amount of virus. The mentioned infectivity parameters contributed an explanation for high levels of the viral load even when the susceptible cells were already depleted at the end of the experiment.

5.3 IN VIVO EBOLA VIRUS INFECTION AND VACCINATION

Experimental observations showed that the immune system often fails to control EBOV infection leading to elevated levels of viral replication [139]. Adaptive immune responses were poor or absent in fatal cases, whereas survivors developed sustained antibody titers [139]. However, follow-up durations were different between fatal cases (approximately one week [140]) and survivors (from a few weeks to months [140]). Currently, treatment of EBOV infection is mainly based on supportive care [141]. Prophylactic and therapeutic approaches are still under development and licensure with promising results for certain antivirals [141–143], passive immunotherapy, and vaccinations [144–146].

EBOV-infected nonhuman primates (NHPs) treated early with monoclonal antibodies (mAbs) were able to recover after challenged with a lethal dose of EBOV [147]. EBOV-infected humans treated with mAbs in addition to intensive supportive care were also more likely to recover [141]. On the other hand, macaques vaccinated early with the rVSV-EBOV vaccine survived lethal EBOV challenge [148]. Based on the rVSV-EBOV vaccine, a recent community trial showed protective efficacy in a ring vaccination trial [149].

These results prompted that the outcome of EBOV infection is sensitive to the time of intervention. Failing to catch up with the infection course could alter the chance to survive EBOV infection. Tailoring time windows of intervention is thus critical at both clinical and epidemiological levels.

Using in vivo experiments in NHPs, Nguyen et al. [57] aimed to model the interactions between EBOV replication and IgG antibody dynamics, with and without passive immunotherapy. In [57], immune response modeling considered only antibody dynamics due to three main reasons. First, antibody responses have shown to be a consistent and long-lasting protective factor in EBOV infection [139,148,149]. Second, when innate immune responses might help in controlling EBOV infection, the effect is transient and debatable [139,150]. Furthermore, taking into account this effect or adaptive cell immune response would only yield a more optimistic estimate. Third, antibody responses have been well reported in highly controlled experimental studies [148,151], facilitating the parameter estimation in modeling processes.

Experimental data. Viral load data in the control and treated cases were extracted from both the studies [147,148]. The IgG antibody dynamics was available to assess its effects on the viral load [148], whereas the mAbs dynamics was only available in terms of administrated time points and dosages. NHPs are considered the best animal model to recapitulate EVD observed in humans [139]. In addition, controlled and defined experimental conditions are key guidance to model the complex interactions between virus and immune responses, for example, defined time of infection and inoculum, consistent sampling time among the subjects, uniform host conditions, and consequently immune responses. Epidemiological and pharmacological studies reported that a viral load higher than 10^6 copies/mL [141] is associated with a higher mortality rate, whereas observations on experimental data in NHPs showed that animals with viral load levels higher than 10^6 TCID$_{50}$ were fatal [148,147]. Thus it is assumed that subjects with a viral load level higher than this threshold will have a *severe outcome*.

Experimental data considering a therapeutic vaccine using monoclonal antibodies (mAbs) was taken from [147]. The mAbs were engineered to specifically recognize the EBOV glycoprotein (GP) inserted in the membrane of the viral particle. In this experiment, a group of 12 macaques was administrated mAbs intravenously at day 3, 6, 9 post infection with a constant dose of with 50 mg/kg. These were divided into two groups; 6 NHPs received the monoclonal antibodies combination ZMapp1 (Group A), and the other 6 received ZMapp2 (Group B) [147]. No treatment was given to the two control cases. EBOV titer increased rapidly but ceased when the first dose of mAbs was administrated. The viral load continued to increase in the

■ **FIGURE 5.8 Experimental designs of the two data sources [57].** The experiments were conducted on nonhuman primates (NHPs). The numbers in the parentheses are the sample size. The days zero indicate the time of Ebola virus infection. *mAbs*: time of monoclonal antibody treatments [147]; *VSV*: time of vaccination with the recombinant vesicular stomatitis virus-Zaire EBOV (rVSV-EBOV) [148].

control cases until the subjects were euthanized at day 7 post-infection. All animals cleared the virus from day 10 onward, with the exception of one animal that presented a high clinical score.

Experimental data for a prophylactic vaccine based on recombinant vesicular stomatitis virus expressing the EBOV GP (VSV-EBOV) was taken from [148]. In this experiment, groups of two or three macaques were vaccinated at 3, 7, 14, 21, and 28 days before EBOV challenge. Macaques were immunized with a single intramuscular injection of plaque-forming units (PFU) of rVSV-EBOV. An ineffective vaccine (the VSV-Marburg virus vaccine, VSV-MARV) was given to three control cases. IgG titers were measured regularly four weeks before and after the challenge. All the vaccinated animals showed a sharp increase of IgG titers one week after vaccination. IgG titers sustained at the level above up to two months. EBOV titers were monitored up to 9 days after the challenge. All the control cases showed a high level of viral titers and were euthanized five to seven days after infection. Viral titer was not observed in all the animals vaccinated at least seven days before the challenge. Among three animals vaccinated three days before the challenge, two had an observable viral load, in which one died, and the other survived. A schematic representation of both NHPs experiments is provided in Fig. 5.8.

EBOV replication in NHPs. EBOV replication dynamics in the absence of any interventions were modeled using EBOV titers (in $TCID_{50}$) of only the control cases in the used data sets [147,148]. Nguyen et al. [57] considered two models, including the logistic growth model and a modified logistic growth model as follows:

$$\text{Logistic}: \quad \frac{dV}{dt} = r_V V \left(1 - \frac{V}{K_V}\right), \tag{5.6}$$

$$\text{Lag-Logistic}: \quad \frac{dV}{dt} = r_V V \left(1 - \frac{V}{K_V}\right)\left(\frac{V}{I_n + V}\right), \tag{5.7}$$

where r_V denotes the virus replication rate, K_V denotes the carrying capacity of the host, and the parameter I_n expresses a threshold below which the virus replication is restrained. Both models assumed that the viral replication is only limited by the available resources of the host.

Table 5.2 Parameter values for Ebola model in vivo and vaccination [57]

Parameters for the viral model (5.7).			
r_V	K_V	I_n	$V(0)$
5.4107	72880400	15.0494	$10^{0.15}$

Parameters for the system (5.8)–(5.11).									
δ_{Ag}	δ_{Ag}	β_{Ag}	β_{Ab}	τ_{Ag}	r_{Ab}	K_{Ab}	$Ag(0)$	$G_{Ag}(0)$	$Ab(0)$
1.1187	0.0248	0	0.0263	3.1574	0.0815	10^4	5×10^7	0	0

The goodness of fit of the models was compared when needed to rule out less supportive models. Considering the AIC, model (5.7) with a lag-phase early after infection and slow growing phase (AIC $= -10$) portrayed the data better than the logistic growth model (AIC $= 21$) in (5.6) (see also Fig. 5.9B and the parameter estimates in Table 5.2). EBOV needed approximately three days to gain the momentum before growing exponentially, suggesting that this is a crucial period for a successful treatment. Noting that the EBOV replication profile represents the cases infecting with a lethal dose, as such a varied subject-specific lag-phase as a function of the inoculum can be expected.

Antibody profile after EBOV vaccination. To obtain a general IgG dynamics in EBOV infection, IgG data [148] after EBOV vaccination and before EBOV infection were extracted. These data show the process from the introduction of EBOV antigen to the secretion of the EBOV-specific IgG antibody without further interventions or infections. Once introduced, the antigen (a live attenuated recombinant vaccine) could be cleared by two main processes: the antigen could be captured by antigen-presenting cells such as macrophages (dendritic cells are reportedly malfunction in EBOV infection [139]) and transported to lymph nodes [152], or by combining with newly synthesized antibodies forming antigen–antibody (AgAb) complexes, which are then also phagocytosed. These dynamics can be written as follows:

$$\frac{dAg}{dt} = -\delta_{Ag} Ag - \beta_{Ag} Ag Ab, \tag{5.8}$$

where δ_{Ag} and β_{Ag} denote the removing rates of the antigen by phagocytic cells and by binding with antibody, respectively. The processed immunogen will then elicit antibody secretion through a series of events. Briefly, naive B cells are activated by the binding with antigen and experienced the formation of the germinal center in the lymph node, leading to B-cell maturation, differentiation, proliferation, and consequently antibody secretion [152]. The time needed for these processes can be summarized as an auxiliary delay state as follows:

$$\frac{dG_{Ag}}{dt} = \frac{\delta_{Ag} Ag - G_{Ag}}{\tau_{Ag}}. \tag{5.9}$$

Nguyen et al. [57] assumed that it takes τ_{Ag} days to process the immunogen to immunogenic signal G_{Ag}. Noting that although antibody production could still be stimulated and remained at high levels for a long time [146], there are experimental evidences showing a quick decaying of IgG titers in two or six months [153]. Thus, to stay on the conservative end, it was assumed in [57] that antibody production will halt when all the antigens are already cleared. Afterward, the B-cell proliferation and antibodies secretion can be summarized as

$$\frac{dAb}{dt} = r_{Ab} G_{Ag} - \beta_{Ab} Ag\, Ab - \delta_{Ab} Ab, \tag{5.10}$$

where r_{Ab} reflects the rate of antibody production, and δ_{Ab} is the natural decaying rate of IgG, which is approximately 28 days [153]. The parameter $\beta_{Ab} = \rho \beta_{Ag}$ denotes the removing rate of antibody from binding with the antigen with $\rho > 0$ reflecting potential varied stoichiometry in the antigen–antibody interaction [154]. Fitting model (5.8)–(5.10) to the average IgG data of all subjects resulted in a classic antibody response picture, namely a lag phase followed by an exponential phase before reaching a plateau (Fig. 5.9A). EBOV-specific IgG antibody appeared negligible during the first week, and a high and steady level can only be acquired two weeks after vaccination.

Tailoring windows of opportunity for prophylactic vaccines. Experimental observations and modeling studies showed that immune responses dynamics are not dependent on viral dynamic but mainly on the initial stimulation of the immune cells. Thus it is assumed that the IgG dynamics elicited from a vaccination is independent from the viral replication dynamic. In addition, experimental data showed that challenging vaccinated NHPs with EBOV boosted IgG dynamics from the immediate state to a higher level [151]. In the used data [148], IgG dynamics showed also a slight boost in some animals after a week.

It is assumed in [57] that EBOV infection would elicit the same IgG dynamic as in the case of vaccination modeled by (5.10), that is, the term Ab in (5.11). Consequently, IgG dynamics promoted from EBOV infection is accumulated to the IgG dynamics elicited from a vaccine supplied to the

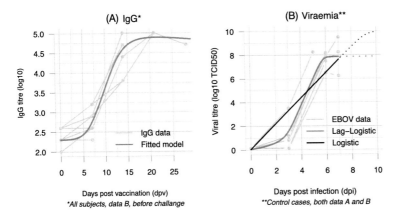

■ **FIGURE 5.9 Fitting models to IgG and viral titer data.** Gray lines are subject-specific data. **(A)** Data of IgG titer post vaccination and prior to EBOV challenge were used to fit to models of antibody responses. The fitted values of the models were superimposed (orange (mid gray in print version) line) illustrating an average profile of IgG dynamics after exposure to EBOV. **(B)** Viral titers in control cases from both data sources [147,148] were used to evaluate EBOV replication models in treatment-free scenarios. The follow-up data were stopped when the animals reached the endpoints to be euthanized [147,148]. The figure is taken from [57].

host κ days prior to the infection, that is, the term $Ab(\kappa + t)$ in (5.11). Considering these altogether, the viral replication model (5.7) was extended to

$$\frac{dV}{dt} = r_V V \left(1 - \frac{V}{K_V}\right)\left(\frac{V}{I_n + V}\right)\left(1 - \frac{Ab + \omega Ab(t + \kappa)}{K_{Ab}}\right). \quad (5.11)$$

The parameter K_{Ab} was introduced to the EBOV replication dynamics (5.7) to reflect a functional threshold at which the antibody titers inhibit EBOV net growth rate. Crossing this threshold leads to the viral clearance. Only subjects vaccinated κ days before the infection have $\omega = 1$; otherwise, $\omega = 0$. To test if this model can reproduce the data, IgG titer and viral load data of two animals vaccinated three days before EBOV infection were used (M31 and M32); these were the only two animals with detectable viral titers in the experiment [148].

Firstly, the antibody dynamics Ab were obtained by fitting equations (5.8)–(5.10) to the IgG data of the two subjects. The parameter r_{Ab} was refitted to allow subject-specific responses while fixing the rest of parameters to the previously obtained estimates from the data of all subjects (Fig. 5.9A). Afterward, the Ab outputs were used as inputs to fit model (5.11) to viral titers data of the two subjects. With the assumption that EBOV replicates similarly among infected subjects, only the parameter K_{Ab} was estimated while fixing the other parameters to the previously obtained estimates from model (5.7).

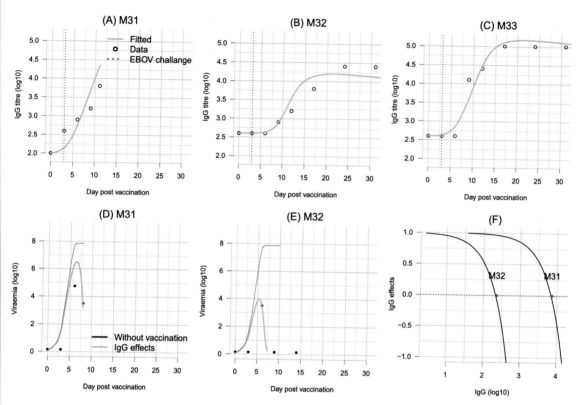

■ FIGURE 5.10 Effects of IgG antibody on controlling viral load. (A–C) Fitting IgG dynamics model (5.8)–(5.10) to IgG data of the three subjects vaccinated three days before EBOV challenge. Note that the M31 had an exceptional low base line IgG (100), whereas all other animals had a higher base line (ranged from 200 to 400 with mean of 269). **(D–E)** Fitting viral dynamics model (5.11) to viral load data of the two subjects vaccinated three days before EBOV challenge. The model without vaccination (solid black line) is added as reference. **(F)** Functions of IgG effect $(1 - \frac{Ab}{K_{Ab}})$ on controlling viral growth in each subject. The figure is taken from [57].

Figs. 5.10D–E show that model (5.11) can reproduce the viral dynamics in the two subjects (M31 and M32). Different estimates of K_{Ab} were obtained for each subject, suggesting possibly subject-specific antibody working threshold. By varying the time of vaccination κ, a time window during which vaccination could prevent a likely lethal viral load level can be estimated. Since the working threshold of antibody response (K_{Ab}) can be subject-specific (Fig. 5.10F), the range of thresholds based on the observed IgG data from $10^{2.5}$ to $10^{4.5}$ were tested.

Based on data of the control cases (Fig. 5.9B) and empirical observations in EBOV-infected human [155], a subject with viral load level higher than 10^6 can be considered as having a severe outcome. Fig. 5.11 shows the time windows for different antibody working thresholds K_{Ab} and vaccination times κ.

For each K_{Ab}, there is a safe time window where viral titers cannot be observed. In fact, the higher the IgG concentration required to suppressed viral replication (K_{Ab}), the narrower the time windows for a successful intervention. For example, the threshold $K_{Ab} = 10^4$ can prevent EBOV replication from reaching severe viral load levels if only a subject had been vaccinated at least 6 days before infection. However, if the subject had received vaccination for more than four months before, the circulating antibody level can decrease below the required working threshold K_{Ab} at the time of infection. As such, if secondary antibody responses to EBOV infection are not considerably faster than primary responses, then the subject would also succumb to the disease.

The secondary antibody response to EBOV infection was assumed similar to a primary response model in vaccination and that the IgG titers were accumulative to the primary response [57]. Remarkably, assuming that an infected subject develops the same IgG profile as a vaccinated subject at the day of the challenge, simulation results showed that the general IgG response fails to keep the viral load from reaching its peak, regardless of the working threshold K_{Ab}. At best, the general IgG profile developing from the day of infection can only clear the virus 9 days post-infection if the subject is still alive after several days withstanding massive viral titers.

5.4 **CONCLUDING REMARKS**

Ebola virus (EBOV) is highly pathogenic for humans, being nowadays one of the most lethal pathogens worldwide. Ebola fatalities are predominantly associated with uncontrolled viremia and lack of an effective immune response (i.e., low levels of antibodies and no cellular infiltrates at sites of infection) [126]. The mathematical model presented by Nguyen et al. [45] focused on the interaction between EBOV and the host cells, that is, epithelial cells of the green monkey. Experimental data on the Vero cell line from nonhuman primates can help to better understand the virus infection dynamics in humans. However, the in vitro studies must be translated carefully to avoid overinterpretation to the in vivo context, which can sometimes lead to erroneous conclusions. The EBOV infection has been known to have abnormal behavior in vivo where different cells types and the immune system are involved.

Nguyen et al. [57] modeled EBOV replication dynamic in situations with and without intervention using NHPs data. These sources of data possess advantages that are not affordable in human settings, the ability to conduct the desired experiment with controlled and known conditions. The left side estimate of the window (Fig. 5.11) appeared to exist when vaccinated-NHPs rechallenged at day 113 ($\approx$ four months), were all survived, whereas vaccinated-NHPs rechallenged at day 234 ($\approx$ eight months) were all died [151]. In humans, there are inconsistent results on the longevity of the antibody when some studies showed sharp drop-offs in antibody titers after two or six months [153], whereas others showed that antibody titers remained

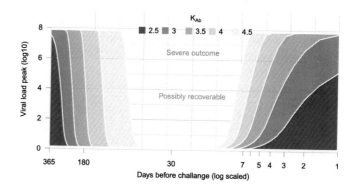

■ **FIGURE 5.11 Simulating windows of opportunity for EBOV vaccination.** Model (5.11) of viral dynamics in the presence of antibody is used. The x-axis shows the time of vaccination from one year to one day before a subject is challenged with a lethal dose of EBOV. Five shades of color represent five assumed antibody working thresholds K_{Ab}: the darker the shade, the stronger the antibody effect in inhibiting the viral replication. Each combination of the vaccination time and the working threshold was used to generate the viral replication dynamics from which its peak was retrieved and plotted in the y-axis. For example, the white area indicates the period when the vaccination completely represses the viral replication with $K_{Ab} = 4.5$. The *severe outcome* indicates the viral load level that was associated with lethal outcome in EBOV infection. The figure is taken from [57].

unchanged up to 360 days [146]. This prolonged existence of antibody can promisingly extend the window's left side, postulating more optimistic opportunities for EBOV prevention. Nonetheless, to ensure a high protection effect, Nguyen et al. [57] aimed to the conservative end by assuming that antibody production will stop once the antigen is cleared.

Antibody responses showed to be a consistent protective factor in EBOV infection [139,148,151]. Antibody dynamics were considered for mathematical models; nevertheless, other parts of the immune system can still play an important role in EBOV infection. There are suggestions of potential effects of innate immune responses on early control in vaccinated subjects [150,139], which current mathematical models do not take into account. Based on the primary assumption to provide worst-case scenarios to EBOV infection [57], taking into account this effect or adaptive cell immune responses would only result in more optimistic scenarios to vaccination. Antibody response data were also the only well-reported part of the immune system in highly controlled experimental studies [148,151]. Experimental data [148,147] showed that in cases of no interventions, mathematical models in [57] exhibited the same dynamic as EBOV titers, that is, very low viral titters during the first three days and then quickly peaked to 10^6–10^7 TCID$_{50}$ on day 6 post-infection.

Simulations in [57] suggested that even if a host effectively produces the general antibody profile (Fig. 5.9A) after infection, the pace of the response is still too slow to counteract EBOV replication, which needs three days to start an exponential growth (Fig. 5.9B). As IgG antibody titers appeared negligible during the first week post vaccination (Fig. 5.9A) and, during this period, the innate immune responses are also largely ineffective [139], the protection to EBOV infection appeared depending on having a high level of antibodies because all animals vaccinated sufficiently early survived [148]. Differences in the race between the EBOV replication and the antibody response highlight the importance of timeliness in EBOV vaccination: the sooner, the better, but might be not too soon. Simulations showed that the window of opportunity for an effective intervention can be limited, ranging from a few days to four months or more, depending on the immune responses strength and on the infective dose. To be on the safe side, a subject infected with a high dose and the general IgG response requires being vaccinated from a week up to four months before the exposure.

Modeling HIV Infection

CONTENTS

6.1 HIV INFECTION

In 1981, reports of a new disease emerged in the USA showing Kaposis sarcoma and other opportunistic infections originating from immunologic abnormalities. The new disease was called Acquired Immunodeficiency Syndrome (AIDS) based on the symptoms, infections, and cancers associated with the deficiency of the immune system. In 1983, Sinoussi and Montagnier isolated a new human T-cell leukemia virus from a patient with AIDS [156], which was later named **HIV**, Human Immunodeficiency Virus. According to UNAIDS estimates for the year 2015, 36 million persons are infected with the HIV worldwide, and there are approximately 2.3 million new infections and 1.6 million AIDS-related deaths that occurred that year [4]. There are at least two types, HIV-1 and HIV-2, which are closely related. Albeit most AIDS worldwide is caused by the most virulent HIV-1, which has been infecting humans in central Africa for far longer than had originally been thought [11].

Like any other virus, HIV does not have the ability to reproduce independently. Hence it must rely on a host cell to replicate. Each virus particle, whose structure is shown in Fig. 6.1, consists of nine genes flanked by long terminal repeat sequences. The three major genes are gag, pol, and env. The gag gene encodes the structural proteins of the viral core, pol encodes the enzymes involved in viral replication and integration, and env encodes the viral envelope glycoproteins. The other six genomes are smaller; Tat and Rev perform regulatory functions that are essential for viral replication, while the remaining four Nef, Vif, Vpra, and Vpu are essential for efficient virus production. HIV expresses 72 glycoprotein projections composed of gp120 and gp41. Gp41 is a transmembrane molecule that crosses the lipid bilayer of the envelope. Gp120 is noncovalently associated with gp41 and serves as the viral receptor for CD4+ T cells.

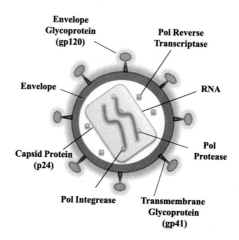

■ **FIGURE 6.1 HIV components.** HIV comprises a viral envelope and associated matrix enclosing a capsid (i.e., protein shell of a virus), which itself encloses two copies of the single-stranded RNA genome and several enzymes. The most important genes and structural proteins are presented.

The HIV genome consists of two copies of RNA, which are associated with two molecules of reverse transcriptase and nucleoid proteins p10, a protease, and an integrase. During infection, the gp120 binds to a CD4 molecule on the surface of the target cell and also to a coreceptor; see Fig. 6.2. This coreceptor can be either the molecule CCR5, primarily found on the surface of macrophages and CD4+ T cells, or the molecule CXCR4, primarily found on the surface of CD4+ T cells. After the binding of gp120 to the receptor and coreceptor, gp41 causes fusion of the viral envelope with the cell membrane, allowing the viral genome and associated viral proteins to enter the cytoplasm.

HIV is classified as a retrovirus, that is, an RNA virus that can replicate in a host cell via the enzyme reverse transcriptase to produce DNA from its RNA genome. The DNA is then incorporated into the cell nucleus by an integrase enzyme. Once integrated, the viral DNA is called a provirus. Then the DNA hijacks the host cell and directs the cell to produce multiple copies of viral RNA. These viral RNA are translated into viral proteins to be packaged with other enzymes that are necessary for viral replication. An immature viral particle is formed, which undergoes a maturation process. The enzyme protease facilitates maturation by cutting the protein chain into individuals proteins that are required for the production of new viruses. The virus thereafter replicates as part of the host cell DNA [157].

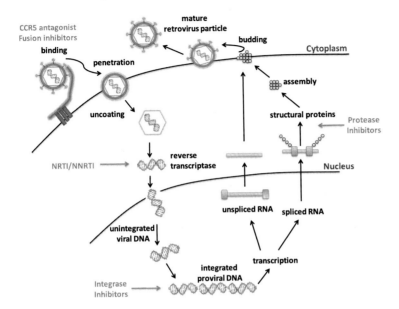

■ **FIGURE 6.2 HIV cycle.** The main parts of HIV cycle are binding, fusion, reverse transcription, integration, replication, assembly, and budding. HIV drugs, antiretrovirals, are grouped into different drug classes according to how they affect HIV replication. HIV medicines approved by the U.S. Food and Drug Administration (FDA) are the Nucleoside Reverse Transcriptase Inhibitors (NRTIs), Non-Nucleoside Reverse Transcriptase Inhibitors (NNRTIs), Protease Inhibitors (PIs), Fusion Inhibitors, and Integrase Inhibitors.

6.2 HIV DISEASE PROGRESSION

The term viral tropism refers to cell types that HIV infects. When a person is infected with HIV, its target is CD4+ T cells, macrophages, and dendritic cells. Because of the important role of these cells in the immune system, HIV can provoke devastating effects on the patients health. For clinicians, the key markers of the disease progression are CD4+ T cell count and viral levels in the plasma.

A typical patient's response consists of an early peak in the viral load, a long asymptomatic period, and a final increase in viral load with simultaneous collapse in healthy T cell count during which AIDS appears. HIV primary infection and disease progression is qualitatively portrayed in Fig. 6.3.

During the **acute infection period** (2–10 weeks), there are a sharp drop in the concentration of circulating CD4+ T cells and a large spike in the

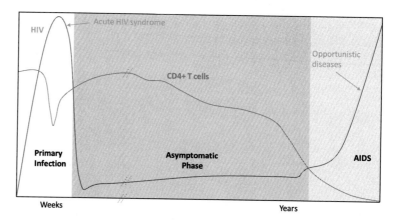

■ **FIGURE 6.3 Typical HIV course.** This scheme shows qualitatively the main patient's responses to HIV infection: an early peak in the viral load, a long asymptomatic phase with a viral quasi-steady state, and a final explosion in viral load accompanied with a drop in CD4+ T cells.

level of circulating free virus (to an average of 10^7 copies/ml). In this primary period, patients developed an acute syndrome characterized by flu-like symptoms of fever, malaise, lymphadenopathy, pharyngitis, headache, and some rash. Following primary infection, seroconversion occurs when people develop antibodies to HIV, which can take from 1 week to several months. After this period, the level of circulating CD4+ T cells returns to near-normal, and the viral load drops dramatically (to an average of about 50,000/ml). In the **asymptomatic or latency period**, without symptoms, the patient does not exhibit any evidence of disease, even though HIV is continuously infecting new cells and actively replicating. The latent period varies in length from one individual to another; there are reports of this latent period lasting only 2 years and more than 15 years as well [158]. Normally, this period ranges from 7 to 10 years. After a long asymptomatic period, the virus eventually gets out of control, and the remaining cells are destroyed. When the CD4+ T cell count has dropped lower than 250 cells/mm^3, the individual is said to have **AIDS**. During this stage the patient starts to succumb to opportunistic infections as the depletion of CD4+ T cells leads to severe immune system malfunction.

6.3 HOW DOES HIV CAUSE AIDS?

AIDS is characterized by the gradual depletion of CD4+ T cells from blood. The mechanism by which HIV causes depletion of CD4+ T cells in infected patients remains unknown. Numerous theories have been proposed, but none can fully explain all of the events observed to occur in patients. The most relevant mechanisms are explained further.

Thymic Dysfunction. The thymus is the primary lymphoid organ supplying new lymphocytes to the periphery. Thymopoiesis, the basic production of mature naive T lymphocytes populating the lymphoid system, is most active during the earlier parts of life. However, recent advances in characterizing thymic functions suggest that the adult thymus is still actively engaged in thymopoiesis and exports new T cells to the periphery until 60 years of age [159]. Several works [160,161] have reported that HIV-induced thymic dysfunction, which could influence the rate of disease progression to AIDS, suggesting a crucial role of impaired thymopoiesis in HIV pathogenesis. Moreover, thymic epithelial cells can also be infected, and this in turn can promote intrathymic spread of HIV [162].

The Homing theory. An important mechanism to explain AIDS is homing (a precisely controlled process where T cells in blood normally flow into the lymph system). This process occurs when CD4+ T cells leave the blood, and then abortive infection with HIV induces resting CD4+ T cells to home from the blood to the lymph nodes; see [163,164]. These homing T cells are abortively infected and do not produce HIV mRNA [164]. The normal lymph–blood circulation process is within one or two days [165], but when they enter in the blood, they exhibit accelerated homing back to the lymph node. Once these abortive cells are in the lymph node, half of them are induced to apoptosis by secondary signals through homing receptors (CD62L, CD44, CD11a) as shown in [164]. The few active infected T cells in lymph nodes bind to surrounding T cells (98–99% of which are resting) and induce signals through CD4+ T and/or chemokine coreceptors.

The Dual role of Dendritic cells. In HIV, Dendritic cells (DCs) play a dual role of promoting immunity while also facilitating infection. C-type lectin receptors on the surface of DCs, such as DC-SIGN, can bind HIV-1 envelope gp120 [166]. DCs can internalize and protect viruses, extending the typically short infectious half-life of virus to several days [167]. The progressive alteration of the immune system resulting in the transition to AIDS could be caused by the dysfunction of DCs. During progression, DCs either fail to prime T cells or are actively immune-suppressive, resulting in failure of the immune control; however, the reasons for this dysfunction are unknown. DCs could also be directly or indirectly affected by HIV causing dysfunction due to a lack of CD4+ T cells [168].

Persistent Immune Activation. Cytopathic effects alone cannot fully account for the massive loss of CD4+ T cells, since productively infected cells occupy a small fraction of total CD4+ T cells (typically of the order of 0.02% to 0.2%). Various clinical studies have linked the massive depletion of CD4+ T cells to the wide and persistent immune activation, which seemed to increase with duration of HIV-1 infection [169,170].

According to this theory, the thymus produces enough naive T cells during the first years of life to fight a lifelong battle against various pathogens. Thus, long-lasting overconsumption of naive supplies through persistent immune activation, such as observed during HIV-1 infection, will lead to accelerated depletion of the CD4+ T and CD8+ T cells stock. This effect would be more pronounced if thymic output depends only on age and not on homeostatic demand; however, this hypothesis is still debated [160].

Immune Escape. For lack of the immune control, different hypotheses have been proposed. The best documented one has been immune escape through the generation of mutations in targeted epitopes of the virus. When effective selection pressure is applied, the error-prone reverse transcriptase and high replication rate of HIV-1 allow a rapid replacement of circulating virus by those carrying resistance mutations, as was first observed with administration of potent antiretroviral therapy [171]. Note that escape may occur even through single amino-acid mutation in an epitope (part of an antigen that is recognized by the immune system) at sites essential for MHC binding or T cell receptor recognition.

Reservoirs and Sanctuary Sites. Long-lived reservoirs of HIV-1 are a barrier to effective immune system response and antiretroviral therapy, and an obstacle for strategies aimed at eradicating HIV-1 from the body. Persistent reservoirs may include latently infected cells or sanctuary sites where antiretroviral drug penetrance is compromised. Moreover, the cell type and mechanism of viral latency may be influenced by anatomical location. Recent studies [172,173] have suggested that latently infected resting CD4+ T cells could be one of these long-term reservoir, whereas other studies have been conducted to explore the role of macrophages as an HIV sanctuary [174].

6.4 MATHEMATICAL MODELING OF HIV INFECTION

Disentangling the leading mechanisms of HIV infection is essential for the design of optimal therapeutic strategies. Mathematical models can serve as a framework to interpret data of ongoing clinical trails, to evaluate the long term of new therapeutic interventions, and to tailor future clinical trials. HIV modeling research twisted in a new dimension when the two works of Perelson et al. [175] and Nowak et al. [48] obtained a mathematical interpretation of viral decay data presented in HIV patients treated with anti-HIV drugs.

Since then, modeling HIV infection has been a very active research topic over the past decades. Most mathematical models present a basic relation between CD4+ T cells, infected CD4+ T cells, and virus [31,42,48,175]. In addition, significant efforts were invested to understand HIV disease progression [176–182], viral persistence [173,183–185], and drug resistance [186–190]. Mathematical modeling studies [54,191] considering activated CD8+ T cells or cytotoxic T cells (CTL) have suggested an important function during HIV infection; however, this function is thought to be compromised during the progression to AIDS.

Single compartment models are able to describe the primary infection and the asymptomatic stage of infection. However, they are not able to describe the transition to AIDS. Most of the mathematical models are based on ordinary or partial differential equations, whereas other authors proposed random variations because of the stochastic nature of HIV infection [192, 193]. A few studies characterize the problem as a cellular automata model to study the evolution of HIV [194,195]. These models have the ability to reflect the clinical timing of the evolution of the virus. To obtain a more widely applicable model, some authors have tried to introduce other variables, taking into consideration other mechanisms by which HIV causes depletion of CD4+ T cells.

Numerous theories have been proposed to explain HIV infection, but none can fully explain all events observed to occur in practice [160,166,170–172, 178,196,197]. Recent laboratory studies [163,164,198] have shown that HIV infection promotes apoptosis in resting CD4+ T cells by the homing process, that is, abortive infection of resting CD4+ T cells may induce to home from the blood to the lymph nodes. This mechanism was modeled in two compartments by Kirschner et al. [199], and simulation results showed that therapeutic approaches involving inhibition of viral-induced homing and homing-induced apoptosis may prove beneficial for HIV patients. Several other investigators have reported that HIV induces thymic dysfunction, which could influence the rate of the disease progression to AIDS. In fact, infection of the thymus could act as a source of both infectious virus and infected CD4+ T cells [200].

Dendritic cell interactions were analyzed and described in a mathematical model in [178], and two main hypotheses for the role of DC dysfunction in progression to AIDS were proposed. The first hypothesis suggests that CD4+ T cells may become depleted by HIV infection, leading into insufficient numbers to license DC, which in turn reduces the ability of DC to prime CD8+ T cells. The second hypothesis suggests that DC dysfunction is the result of a direct viral effect on DC intracellular processes.

6.5 THE THREE STAGES IN HIV INFECTION

A reservoir is a long-lived cell, which can have viral replication even after many years of drug treatment. Experimental studies [172,173] have suggested that CD4+ T cells could be one of the major viral reservoirs. HIV replicates well in activated CD4+ T cells, and latent infection is thought to occur only in resting CD4+ T cells. Latently infected resting CD4+ T cells provide a mechanism for life-long persistence of replication-competent forms of HIV, rendering hopes of virus eradication with current antiretroviral regimens unrealistic. However, experimental observations [201] revealed that the virus reappearing in the plasma of patients undergoing interruption of a successful antiviral therapy is genetically different from that harbored in latently infected CD4+ T cells by HIV. These strongly suggest that other reservoirs may also be involved in the rebound of HIV replication.

A number of clinical studies have been conducted to explore the role of macrophages in HIV infection [174]. Macrophages play a key role in HIV disease; they appear to be the first cells infected by HIV, to spread infection to the brain, and to form a long-lived virus reservoir. A mathematical model that describes the complete HIV/AIDS trajectory was proposed in [180]. Simulation results for that model emphasize the importance of macrophages in HIV infection and progression to AIDS. Hadjiandreou et al. [180] developed a good mathematical model to describe the whole HIV infection course; however, further work is needed since the model is very sensitive to parameter variations.

A simplification from [180] is performed with the following populations: T represents the uninfected CD4+ T cells, T^* represents the infected CD4+ T cells, M represents uninfected macrophages, M^* represents the infected macrophages, and V represents the HIV population. The mechanisms considered for this model are described in [182] by the following reactions.

Homeostatic cell proliferation. The source of new CD4+ T cells and macrophages from thymus, bone marrow, and other cell sources is assumed to be constant [202]:

$$\varnothing \xrightarrow{s_T} T, \tag{6.1}$$

$$\varnothing \xrightarrow{s_M} M, \tag{6.2}$$

where s_T and s_M are the source terms and represent the generation rate of new CD4+ T cells and macrophages, which were estimated as 10 cells/mm^3 day and 0.15 cells/mm^3 day, respectively, by [202].

Natural death. Cells and virus have a finite lifespan. This loss is represented by the following reactions:

$$T \xrightarrow{\delta_T} \varnothing, \tag{6.3}$$

$$T^* \xrightarrow{\delta_{T^*}} \varnothing, \tag{6.4}$$

$$M \xrightarrow{\delta_M} \emptyset, \tag{6.5}$$

$$M^* \xrightarrow{\delta_{M^*}} \emptyset, \tag{6.6}$$

$$V \xrightarrow{\delta_V} \emptyset. \tag{6.7}$$

The death rate of CD4+ T cells in humans is not well characterized; this parameter has been chosen in a number of works as $\delta_T = 0.01$ day^{-1}, a value derived from BrdU labeling macaques [202]. The infected cells were taken from [202] with values of δ_{T^*} ranging from 0.26 to 0.68 day^{-1}; this value is bigger than uninfected CD4+ T cells because infected CD4+ T cells can be cleared by CTL cells and other natural responses. In contrast to CD4+ T cells, HIV infection is not cytopathic for macrophages, and the half-life of infected macrophages may be of the order of months to years depending on the type of macrophage. These long-lived cells could facilitate the ability of the virus to persist [203]. Moreover, studies of macrophages infected in vitro with HIV showed that they may form multinucleated cells that could reach large sizes before degeneration and necrosis ensued [174]. The current consensus is that the principal cellular target for HIV in the CNS (Central Nervous System) is the macrophage or microglial cell. A large study in clinical well-characterized adults found no convincing evidence for HIV DNA in neurons [204]. Thus macrophages and infected macrophages could last for very long periods; δ_M and δ_{M^*} are estimated as 1×10^{-3} day^{-1} using clinical data for the CD4+ T cells [205,206]. Clearance of free virions is the most rapid process, occurring on a time scale of hours. The values of δ_V ranged from 2.06 to 3.81 day^{-1} [202,207].

Infection process. HIV can infect a number of different cells: activated CD4+ T cells, resting CD4+ T cells, quiescent CD4+ T cells, macrophages, and dendritic cells. For simplicity, just activated CD4+ T cells and macrophages are considered in the infection process:

$$T + V \xrightarrow{k_T} T^*, \tag{6.8}$$

$$M + V \xrightarrow{k_M} M^*. \tag{6.9}$$

The parameter k_T is the rate at which a free virus V infects CD4+ T cells; this has been estimated by different authors, and the range for this parameter is from 10^{-8} to 10^{-2} ml/day copies [180]. The macrophage infection rate k_M is fitted as 2.4667×10^{-7} ml/day copies.

Virus replication. HIV may be separated into their source, either CD4+ T cells or macrophages, by the host proteins contained within their coat [208]. Viral replication is considered as occurring in activated CD4+ T cells and macrophages:

$$T^* \xrightarrow{p_T} V + T^*, \tag{6.10}$$

$$M^* \xrightarrow{p_M} V + M^*. \tag{6.11}$$

The amounts of virus produced from infected CD4+ T cells and macrophages are given by $p_T T^*$ and $p_M M^*$, respectively, where p_T and p_M are the rates of production per unit time in CD4+ T cells and macrophages. The values for these parameters are in a very broad range depending on the model, cells, and mechanisms. We take the values from [180], where p_T ranges from 0.24 to 500 copies mm^3/cells ml, and from 0.05 to 300 copies/cells day for p_M. Note that not all virus particles are infectious; only a limited fraction ($\approx 0.1\%$) of circulating virions are demonstrably infectious [168]. Some virus particles have defective proviral RNA, and therefore they are not capable of infecting cells. In mathematical models, generally, V describes the population dynamics of free infectious virus particles.

Model Simulation. Assuming the mechanisms (6.1)–(6.7) as the most relevant, we obtain the following mathematical equations:

$$
\begin{aligned}
\dot{T} &= s_T - k_T T V - \delta_T T, \\
\dot{T}^* &= k_T T V - \delta_{T^*} T^*, \\
\dot{M} &= s_M - k_M M V - \delta_M M, \\
\dot{M}^* &= k_M M V - \delta_{M^*} M^*, \\
\dot{V} &= p_T T^* + p_M M^* - \delta_V V.
\end{aligned}
\tag{6.12}
$$

For initial model condition values, previous works are considered [202, 180]: CD4+ T cells are taken as 1000 cells/mm^3 and 150 cells/mm^3 for macrophages. Infected cells are considered as zero and initial viral concentration as 10^{-3} copies/ml. In this chapter, as a learning process, parameter fitting was realized by a trial-and-error process to match clinical data in [205,206].

Considering parameter values presented in Table 6.1, numerical results given in Fig. 6.4 show a fast drop in healthy CD4+ T cells, whereas there is a rapid increase in viral load. It might be expected that the immune system responds to the infection, proliferating more CD4+ T cells, which gives rise to the increment in CD4+ T cells. However, in (6.12) there is no term for proliferation, and therefore the observed increase in CD4+ T cell count is due to a saturation of infection in CD4+ T cells and a consequent sharp drop in the viral load experienced. For approximately 4 to 5 years, an untreated

Table 6.1 Parameters values for the mathematical model represented by (6.12)

Parameter	Nominal Value	Reference	Parameter Variation
s_T	10	[202]	7–20
s_M	0.15	[202]	0.1–0.3
k_T	3.5714×10^{-5}	[180]	3.2×10^{-5}–1.0×10^{-4}
k_M	4.3333×10^{-7}	Adjusted	3.03×10^{-7}–1.30×10^{-6}
p_T	38	[180]	30.4–114
p_M	44	[180]	22–132
δ_T	0.01	[180]	0.001–0.017
δ_T^*	0.4	[202]	0.1–0.45
δ_M	1×10^{-3}	Adjusted	1×10^{-4}–1.4×10^{-3}
δ_M^*	1×10^{-3}	Adjusted	1×10^{-4}–1.2×10^{-3}
δ_V	2.4	[180]	0.96–2.64

patient experiences an asymptomatic phase where in CD4+ T cell counts levels are over 300 cells/mm^3. On one hand, CD4+ T cells experience a slow but constant depletion; on the other hand, the virus continues infecting healthy cells, and therefore a slow increase in viral load takes place, as can be seen in Fig. 6.4.

At the end of the asymptomatic period, HIV symptoms appear when CD4+ T cell counts are below 300 cells/mm^3. The last stage and the most dangerous for the patient is when the depletion in CD4+ T cells crosses 250 cells/mm^3, which is considered as AIDS. This is usually accompanied by a rapid growth in viral load, and the severe immuno-deficiency frequently leads to potentially fatal opportunistic diseases. Fig. 6.4 reveals how the model is able to represent the three stages in HIV infection and corresponds reasonably well to clinical data.

Macrophages are considered one of the first points of infection, and therefore infected macrophages may become long-lived virus reservoirs as stated in [174]. Fig. 6.4 shows healthy macrophage dynamics with a slow depletion in counts; this depletion is because of their change to infected status. The number of infected macrophages increases slowly during the asymptomatic period, but when constitutional symptoms appear, infected macrophages increase in population faster than before. These results suggest that in the last stages of HIV the major viral replication comes from infected macrophages. This is consistent with the work of [209], which states that in the early infection the virus replication rate in macrophages is slower than the replication rate in CD4+ T cells. Over the years, the viral replication rate in macrophages grows.

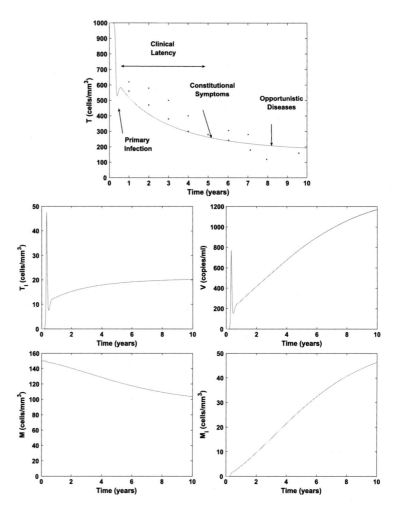

■ **FIGURE 6.4 Dynamics of immune cells and virus over a period of ten years.** Data taken from [205] are represented by □, and data from [206] are represented by ◇. Taken from [182].

Whilst the model in [180] reproduces known long-term behavior, bifurcation analysis gives an unusually high sensitivity to parameter variations. For instance, small relative changes in infection rates for macrophages give bifurcation to a qualitatively different behavior. Therefore it is necessary to check the sensitivity to parameter variation in the proposed model (6.12). Accordingly, parameters can be varied to observe the range for which the model (6.12) shows the whole HIV infection trajectory with reasonable different time scales. For instance, Fig. 6.5A reveals that the model may reproduce long-term behavior despite high variations of the parameter k_T, which may range from 10% below nominal values and 220% above. It can be noticed that higher infection of CD4+ T cells speeds up the progression to AIDS. Table 6.1 reveals that parameters can be varied in a wide range whilst still showing the three stages in HIV infection with reasonable time scales. In this case, it is defined as reasonable time scales for progression to AIDS in between 1 and 20 years. It is considered that (6.12) might be a useful model to represent the whole HIV infection for different patients as a result of its robustness to represent the three stages of HIV infection.

Considering, for instance, the death rate of healthy CD4+ T cells d_T, the initial thoughts may be that increasing the death rate of CD4+ T cells will hasten the progression to AIDS. Nonetheless, Fig. 6.5B provides interesting insights of the progression to AIDS. For instance, if the death rate of CD4+ T cells is small, then the progression to AIDS is faster since CD4+ T cells live for longer periods and become infected, and then more viruses are produced. Moreover, more infection of long-term reservoirs takes place. On the other hand, if the death rate of CD4+ T cells is high, then the viral load explosion might be inhibited.

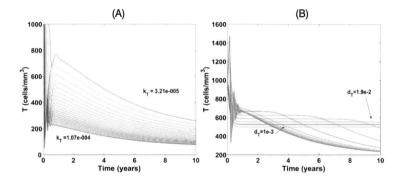

■ **FIGURE 6.5 CD4+ T cell dynamics under parameter variation.** The parameters k_T and d_T were varied individually keeping the other constants. Taken from [182].

Undoubtedly, CD4+ T cells levels will be low with high d_T value, but Fig. 6.5 exposed that there is a range for d_T in which CD4+ T cells could be maintained in safety levels (> 350 cells). Clinical evidence has shown that HIV affects the life cycle of CD4+ T cells [164]. For simplicity, in (6.12) one compartment of activated CD4+ T cells is considered, which are directly infected by HIV. Let us consider d_T as a regulation between two pools of cells, naive and activated cells; consequently; it could be inferred that, for a stronger activation of CD4+ T cells, the progression to AIDS would be faster. Clinical observations [170] have supported the hypothesis that persistent hyperactivation of the immune system may lead to erosion of the naive CD4+ T cells pool and CD4+ T cell depletion. Numerical results yield the idea that macrophages need the first stage of viral explosion to attain large numbers of long-lived reservoirs and cause the progression to AIDS. Thus, a regulation in the activation of CD4+ T cells in the early stages of HIV infection might be important to control the infection and its progression to AIDS.

Steady-State Analysis. Using system (6.12), the equilibria may be obtained analytically in the following form:

$$T = \frac{s_T}{k_T V + \delta_T}, \quad T^* = \frac{k_T s_T}{\delta_{T^*}} \frac{V}{k_T V + \delta_T},$$

$$M = \frac{s_M}{k_M V + \delta_M}, \quad M^* = \frac{k_M s_M}{\delta_{M^*}} \frac{V}{k_M V + \delta_M},$$

where V is the solution of the polynomial equation

$$a V^3 + b V^2 + c V = 0. \tag{6.13}$$

Eq. (6.13) has three solutions

$$V^{(A)} = 0, \quad V^{(B)} = \frac{-b + \sqrt{b^2 - 4ac}}{2a}, \quad V^{(C)} = \frac{-b - \sqrt{b^2 - 4ac}}{2a} \tag{6.14}$$

with

$$a = k_T k_M \delta_{T^*} \delta_{M^*} \delta_V,$$

$$b = k_T \delta_{T^*} \delta_M \delta_{M^*} \delta_V + k_M \delta_T \delta_{T^*} \delta_{M^*} \delta_V - s_T k_T k_M p_T \delta_{M^*} - s_2 k_T k_M p_M \delta_{T^*},$$

$$c = \delta_T \delta_{T^*} \delta_M \delta_{M^*} \delta_V - s_T k_T p_T \delta_M \delta_{M^*} - s_M k_M p_M \delta_T \delta_{T^*}.$$

Equilibrium A:

$$T^{(A)} = \frac{s_T}{\delta_T}, \quad T^{*(A)} = 0, \quad M^{(A)} = \frac{s_M}{\delta_M}, \quad M^{*(A)} = 0, \quad V^{(A)} = 0.$$

Equilibrium B, C:

$$T^{(B,C)} = \frac{s_T}{k_T V^{(B,C)} + \delta_T}, \quad T^{*(B,C)} = \frac{k_T s_T}{\delta_{T^*}} \frac{V^{(B,C)}}{k_T V^{(B,C)} + \delta_T},$$

$$M^{(B,C)} = \frac{s_M}{k_M V^{(B,C)} + \delta_M}, \quad M^{*(B,C)} = \frac{k_M s_M}{\delta_{M^*}} \frac{V^{(B,C)}}{k_M V^{(B,C)} + \delta_M}.$$

Proposition 6.1. *The compact set* $\Gamma = \{\{T, T_i, M, M_i, V\} \in R_+^5 : T(t) \leq s_T/\delta_T, M(t) \leq s_M/\delta_M\}$ *is a positive invariant set.*

Proof. The proof can be found in [210]. $\qquad\square$

Remark 6.1. If c is a negative real number, then (6.12) has a unique infected equilibrium in the first orthant.

Proof. This can be seen directly from $b^2 - 4ac > 0$ in (6.14).   □

Remark 6.2. It is possible to have two infected equilibria in the first orthant if b is a negative real number and c is a positive real number. By using the values from Table 6.1 it can be easily shown that b and c are negative.

Proposition 6.2. *The uninfected equilibrium is locally unstable if there exists a unique infected equilibrium in the first orthant.*

Proof. If the characteristic polynomial is computed for the equilibrium A (uninfected status), then a polynomial of fifth order can be obtained, where one of the coefficients is equal to c. The Routh–Hurwitz criterion yields that the equilibrium is stable if and only if every coefficient of the characteristic polynomial is positive. However, it was previously shown that c must be negative in order to have a unique equilibrium point. Therefore the uninfected equilibrium is unstable, which is consistent with previous works [202,211]. This can be interpreted as why it is impossible to revert a patient once infected back to an HIV-free state.   □

Models with different cell proliferation terms. The macrophage dynamics modeled in this work present a very different scenario from [180], which presented a model with an explosion in macrophage populations (both infected and healthy over 1000 cells/mm^3). Simulation results reveal slightly depletion in uninfected macrophages and increment in infected macrophages, as can be seen in Fig. 6.4, but the total population remains almost constant. This difference in macrophage dynamics is because here so far cell proliferation terms were not considered. To adjust the fast depletion in CD4+ T cells and explosion in viral load, cell proliferation terms are included as follows:

$$T + V \xrightarrow{\rho_T} T + (T + V), \qquad (6.15)$$

$$M + V \xrightarrow{\rho_M} M + (M + V). \qquad (6.16)$$

Using the proliferation rates (6.15) and (6.16) in the proposed model (6.12), the dynamics match better in the final depletion of CD4+ T cells as can be noticed in Fig. 6.6. The collapsing in CD4+ T cells is obvious using the proliferation rates ρ_T and ρ_M. Moreover, after the primary infection stage takes place, the recovering in CD4+ T cell counts over 500 cells/mm^3 is more evident using cell proliferations terms.

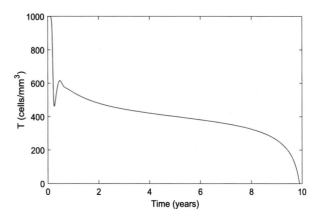

■ **FIGURE 6.6 CD4+ T cells dynamics.** Using bilinear proliferation rates (6.15) and (6.16), the slow depletion of CD4+ T cells in HIV infection can be obtained.

Hadjiandreou et al. [180] emphasized the importance of macrophages in the progression of HIV; however, they had not provided any mathematical evidence of the HIV/AIDS transition. Since the mathematical model is difficult to analyze, to allow mathematical analysis, simplifying assumptions are used based on the following remarks.

Assumption 6.1. Fast Viral Dynamics. Using the parameter values in Table 6.1, we can observe that $\delta_V \gg 1$. This corresponds to viral dynamics with a time constant much less than one day. In this case, the differential equation for the virus can be approximated by the following algebraic equation as suggested in [212]:

$$V = \frac{p_T}{\delta_V} T^* + \frac{p_M}{\delta_V} M^*. \tag{6.17}$$

Assumption 6.2. T^* **is approximately constant during the asymptomatic phase.** We note that in the asymptomatic period of infection (that is, after the initial transient and before the final divergence associated with development of AIDS), the concentration of infected CD4+ T cells is relatively constant. This assumption is also proposed in [211]. Therefore the following approximation for infected CD4+ T cells can be written as follows:

$$T^*(t) \approx \overline{T}^* \ \forall t \geq t_0. \tag{6.18}$$

Using Eqs. (6.17) and (6.18), we derive the following expression:

$$V(t) := c_1 M^* + V_{T^*}, \tag{6.19}$$

where $V_{T*} = \frac{p_T}{\delta_V}\overline{T}^*$ and $c_1 = \frac{p_M}{\delta_V}$. Note that if $\overline{T}^*$ is selected as an upper bound on T^*, then (6.19) represents an upper bound on $V(t)$. Therefore (6.19) describes the long asymptomatic period in the viral load dynamic. Considering now Assumptions (6.1) and (6.2) in (6.12), we can obtain the following equations:

$$\dot{M} \approx s_M - c_2 M + c_3 M M^*, \qquad (6.20)$$

$$\dot{M}^* \approx c_4 M + c_5 M M^* - \delta_4 M^*, \qquad (6.21)$$

where $c_2 = \delta_M - (\rho_M - k_M)V_{T*}$, $c_3 = (\rho_M - k_M)c_1$, $c_4 = k_M V_{\overline{T}*}$, and $c_5 = k_M c_1$.

Assumption 6.3. M **and** M^* **have an affine relation.** Note that from (6.20) and (6.21) it is expected that the bilinear terms are predominant for large M and M^*; then we can assume that $\dot{M} \approx \frac{c_3}{c_5}\dot{M}^*$, which is rearranged in the linear form

$$M^* \approx c_6 M - c_7, \qquad (6.22)$$

where $c_6 = \frac{c_2 c_5 + c_3 c_4}{c_3 \delta_{M*}}$ and $c_7 = \frac{c_5 s_M}{c_3 \delta_{M*}}$.

Proposition 6.3. *Under Assumptions 6.1–6.3 and the parameter condition* $4\alpha s_M \geq \beta^2$, *the macrophage dynamics in HIV infection is unstable with finite escape time.*

Proof. Substituting (6.22) into Eq. (6.20), we obtain a numerically verified approximation for the macrophage equation only for large amounts of M and M_i:

$$\dot{M} = s_M + \alpha M^2 + \beta M, \qquad (6.23)$$

where

$$\alpha = \frac{c_6 p_M (\rho_M - k_M)}{\delta_V},$$

$$\beta = \frac{(\rho_M - k_M)(p_T \overline{T}^* - c_7 p_M)}{\delta_V} - \delta_M$$

The solution of the differential equation (6.23) is given by

$$M = \frac{\beta}{2\alpha} + \frac{\sqrt{4\alpha s_M - \beta^2}}{2\alpha} \tan\left(\frac{\sqrt{4\alpha s_M - \beta^2}}{2}t + \eta\right), \qquad (6.24)$$

where η is a constant related to the initial condition of the macrophages given by

$$\eta = \tan^{-1}\left(\frac{2\alpha M_0 - \beta}{\sqrt{4\alpha s_2 - \beta^2}}\right). \tag{6.25}$$

In (6.24) there is a tangent function if $4\alpha s_M \geq \beta^2$, which tends to ∞ as the argument tends to $\pi/2$, that is, when

$$t = T_\infty := \frac{\pi - 2\eta}{\sqrt{4\alpha s_M - \beta^2}}, \tag{6.26}$$

which implies that there is a finite escape time. $\qquad \square$

Whilst the incorporation of proliferations rates (6.15) and (6.16) in model (6.12) reproduces the observed long-term behavior more accurately, any small change in parameters (i.e., 1% nominal value) evidences an unusually high sensitivity. In particular, small relative changes in k_T, ρ_T, ρ_M, δ_T, δ_{T^*}, δ_M, or δ_V give bifurcation to a qualitatively different behavior. This sensitivity to parameter variation is caused by the unstable behavior of system (6.12) shown in Proposition 6.3. This finite time escape is because the macrophage proliferation rate is faster than the infection rate of macrophages, that is, $\rho_M > k_M$.

In order to have the same trajectory presented in Fig. 6.6 and robustness to parameter variation, the proliferation terms (6.15) and (6.16) are changed using Michaelis–Menten kinetics in the following form:

$$T + V \xrightarrow{\frac{\rho_T}{C_T + V}} T + (T + V), \tag{6.27}$$

$$M + V \xrightarrow{\frac{\rho_M}{C_M + V}} M + (M + V). \tag{6.28}$$

The new parameters were adjusted to obtain the appropriate HIV trajectory with respect to clinical observations. These parameter values are $\rho_T = 0.01$, $\rho_M = 0.004$, $C_T = 300$, and $C_M = 500$. Adding cell proliferation rates (6.27) and (6.28) in model (6.12), the new proposed mathematical model writes as follows:

$$
\begin{aligned}
\dot{T} &= s_T + \frac{\rho_T V}{C_T + V}T - k_T TV - \delta_T T, \\
\dot{T^*} &= k_T TV - \delta_{T^*} T^*, \\
\dot{M} &= s_M + \frac{\rho_M V}{C_M + V}M - k_M MV - \delta_M M, \\
\dot{M^*} &= k_M MV - \delta_{M^*} M^*, \\
\dot{V} &= p_T T^* + p_M M^* - \delta_V V.
\end{aligned}
\tag{6.29}
$$

6.10 SIMULATING HIV MODEL (6.29)

```
clc; clear all; close all;
% Simulation time in years
Tmax=10*365;

% Solving Differential Equations
[t,y]=ode45(@model,[0 Tmax],[1000 0 150 0 1e-3]);% ODE function
T=y(:,1);      % Healthy (uninfected) cells
Ti=y(:,2);     % Infected cells (Active)
M=y(:,3);      % Macrophages
Mi=y(:,4);     % Infected Macrophages
V=y(:,5);      % Virus

% Plotting
figure(1); clf;
sim_V=plot(t/365,T,'LineWidth',2);
axis([0,Tmax/365,1,1000]);
xlabel('Time (days)');
ylabel('T(cells/mm^3)','fontsize',10.0)
hold on
figure(2); clf;
plot(t/365,V/0.001,'LineWidth',2);ylabel('V (copies/mm^3)')
xlabel('Time (years)');
axis([0,Tmax/365,1,2500000]);
figure(3); clf;
plot(t/365,Ti,'LineWidth',2);ylabel('T^* (cells/mm^3)')
xlabel('Time (years)');
figure(4); clf;
plot(t/365,M,t/365,Mi,'LineWidth',2); ylabel('M (cells/mm^3)')
xlabel('Time (years)');

function  dy= model(t,y)
s1=10;s2=0.15;
K2=4.5714e-005;K4=4.3333e-008;
K5=38;K6=35;
d1=0.01;d2=0.4;d3=0.001;d4=0.001;d5=2.4;
pT=0.01*(y(1)*y(5))/(y(5)+300);
pM=0.003*(y(3)*y(5))/(y(5)+220);
dy = zeros(5,1);
dy(1)  = s1 -d1*y(1)-K2*y(5)*y(1)+pT;
dy(2)  = K2*y(5)*y(1)-d2*y(2);
dy(3)  = s2 -K4*y(3)*y(5)-d3*y(3)+pM;
dy(4)  = K4*y(5)*y(3)-d4*y(4);
dy(5)  = K5*y(2)+ K6*y(4)-d5*y(5);
end
```

Box 6.10 contains the Matlab code to simulate model (6.29). Simulations in Fig. 6.7 show how in the symptomatic period CD4+ T cell counts drop dramatically lower than 250 cell/mm^3. In addition, the primary infection dynamics are adapted better to clinical observations, that is, when CD4+ T cells drops below 400 (cells/mm^3) and then returns to near-normal values. Viral load trajectories agree with clinical observations; a large spike in the level of circulating virus followed by a fast drop in viral concentration. In the latent period the viral load remains almost constant, and finally the explosion in viral load takes place in the symptomatic stage. The relevance of these cell proliferation rates incorporated in model (6.29) is that robust properties to parameter variation are preserved and can be varied as shown in Table 6.1. Viral explosion promotes more infection of long-term cells, which will replicate virus during long periods. This is consistent with simulation results in [180], which found that infected macrophages increase slowly in number during the asymptomatic period and exponentially in the later stage of the disease.

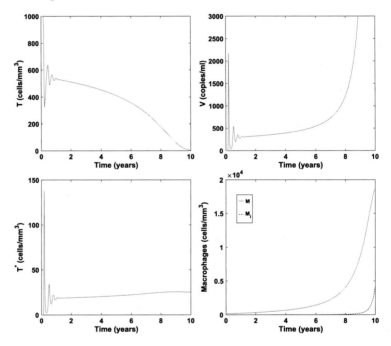

■ **FIGURE 6.7** CD4+ T cell, macrophage and viral dynamics using proliferation rates (6.27) and (6.28).

Macrophages together with Langerhans cells are considered one of the first points of infection, and therefore infected macrophages may become long-lived viral reservoirs as stated in [174]. Immune system response promotes the proliferation of different cell lines; for this reason, we can note in Fig. 6.7 a logistic growth in macrophages. The number of infected macrophages increases slowly but consistently over time; see Fig. 6.7. These results suggest that in the last stages of HIV the majority of viral replication comes from infected macrophages. This is consistent with the work [209], which states that, during early infection, the viral replication rate in macrophages is less than the replication rate in CD4+ T cells.

There is lack of information regarding infected cells in HIV. Simulation results suggest that infected macrophages may experience an increase in population as can be seen in Fig. 6.7. This is consistent with studies in rhesus macaques [209] using the highly pathogenic simian immunodeficiency virus/HIV type 1 (SHIV). This is an exaggerated model of HIV infection in humans that allows scientist to address certain clinical aspects of retrovirus that are difficult to study in people. Lymphoid organs such as lymph nodes and spleen for the source of the remaining virus were examined in [209]. They found that 95% of the virus-producing cells were macrophages and only 1 to 2% were CD4+ T cells. Moreover, macrophages contain and continue to produce large amounts of HIV-like virus in monkeys even after CD4+ T cells are depleted.

6.6 **CONCLUDING REMARKS**

Mathematical modeling helped to elucidate two main mechanism, which can be considered as two feedback systems. One provides the fast dynamics presented in the early stages of infection as a result of an strong inhibition to CD4+ T cells. The second feedback sustains a constant slow infection process in macrophages over the years due to a weak inhibition accompanied by the long-time survival conditions of macrophages; see Fig. 6.8.

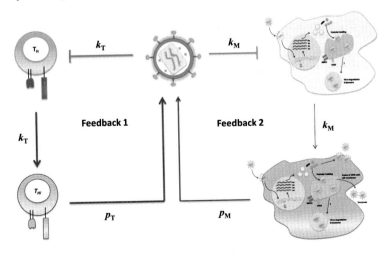

■ **FIGURE 6.8 HIV infection scheme.** The mechanisms to explain the three stages of HIV infection can be described with two groups of cells.

We can distinguish the different contributions to the viral load in Fig. 6.9. The viral load of infected CD4+ T cells would result in very fast convergence, showing that the final viral load explosion is not due to infected

CD4+ T cells; this is also known from previous works [202]. In this model, the final viral explosion is due to infected macrophages; nevertheless, they would be unable to generate the initial peak in viral load. Notice that macrophage population in (6.29) is essential for explaining the progression to AIDS. Nonetheless, other cell populations or reservoirs can also explain the dynamics. There appear to be two key factors in these dynamics: (i) the reservoir should have very long lives and slow background proliferation; (ii) the reservoir dynamics should be largely decoupled from the CD4+ T cell dynamics.

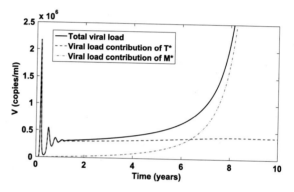

■ FIGURE 6.9 Viral load contribution by the tape of infected cells. Infected CD4+ T cells drive the first stage of HIV dynamics, whereas in the last stage, HIV is mainly driven by infected macrophages. Taken from [182].

The model proposed in [180] exhibits both exponential growth in macrophage population and high parameter sensitivity. Both of these behaviors do not seem to be realistic. In our proposed model (6.29), a very rapid growth occurs if bilinear proliferation terms such as $\rho_M M V$ for macrophages are used. In this case, if the macrophage proliferation rate is faster than the infection rate of macrophages ($\rho_M > k_M$), then the macrophage population grows exponentially fast.

Note that any function with logistic growth, for instance, Michaelis–Menten or Hill functions, can lead to more realistic results. Several models suggest logistic growth for cell proliferation; see, for example, [202]. This means that the proliferation cell process is relevant to represent the adequate dynamics. Another way to obtain realistic behavior is considering the homeostatic proliferation of macrophages population limited by $M + M^*$. This also would allow sigmoidal convergence to a high percentage of the macrophages being infected without explosion in macrophage population.

A Monte Carlo sampling study was developed to check the cross impact of the parameter variations. To this end, 1000 evaluations of the proposed deterministic model (6.29) are performed, using different sets of parameters with normally distributed random variations about the nominal values.

Table 6.2 Joint parameter variation with 1000 simulations

Parameter Variation (%)	Number of Progressors	Average progression time (years)
5	1000	7.15
10	1000	7.37
30	985	7.57
50	868	7.52
80	711	6.95
100	648	6.87

Numerical study in Table 6.2 shows how the proposed model is able to represent the three stages of the infection even under a full set of parameter variations. For instance, nominal parameter variations of up to 10% provide the full dynamics with slightly different time scales with an average disease progression of 7.3 years. Note that clinicians report different time scales for progression to AIDS (3 to 15 years) [205]. Table 6.2 reveals that larger variations may be explored in the model having still a high number of progressors to AIDS. We can conclude that the proposed model allows parameter perturbations of approximately 50%, whilst still reproducing the three stages of infection in more than 80% of the cases. Therefore the proposed model retains its ability to describe the three stages in HIV infection even for moderately large full parameter variations.

Macrophage population is essential for explaining the progression to AIDS. Nonetheless, other classes of reservoirs can also explain this progression. The progression to AIDS in HIV infection is still an open problem for discussion in clinical circles, where further experimental evidence of macrophages could either confirm or falsify the hypothesis presented in [182].

HIV Evolution During Treatment

7.1 ANTIRETROVIRAL DRUGS FOR HIV INFECTION

The most important scientific advance after the identification of HIV as the causative agent for AIDS was the development of effective antiretroviral drugs for treating individuals infected with HIV. The first effective drug against HIV was the reverse transcriptase inhibitor Zidovudine, which was developed as an anticancer drug but was not effective in that capacity. It was licensed as the first antiretroviral drug in 1987. The use of Zidovudine during pregnancy was documented to decrease neonatal transmission of HIV from 25.5% to 8.3% [213]. Thus far, over 25 antiretroviral agents have been developed to target specific vulnerable points in the HIV life cycle; see Fig. 6.2.

A "cocktail" of three or more potent antiretroviral drugs is known either as combination antiretroviral therapy (cART) or highly active antiretroviral therapy (HAART). HIV treatment can reduce viral replication and delay the progression to AIDS [214]. Another salutary effect of HAART is the restoration of immune function, which routinely occurs on long-term therapy and leads to the regeneration of robust CD4+ T and CD8+ T cellular responses to recall antigens [215].

Six different antiretroviral classes are presented in Table 7.1, which are nucleoside/nucleotide reverse transcriptase inhibitors (NRTIs), nonnucleoside reverse transcriptase inhibitors (NNRTIs), protease inhibitors (PIs), fusion inhibitors (FIs), CCR5 antagonist, and integrase strand transfer inhibitors

Modeling and Control of Infectious Diseases in the Host. https://doi.org/10.1016/B978-0-12-813052-0.00018-X

(INSTI). The most extensively studied combination regimens for treatment-naive patients that provide durable viral suppression generally consist of two NRTIs plus one NNRTI or PI [216].

Table 7.1 Antiretroviral Drugs [216]

Class	Generic Name	Trade Name	Intracellular Half-life
NNRTIs	Delavirdine	Rescriptor	5.8 hrs
	Efavirenz	Sustiva	40–55 hrs
	Etravirine	Intelence	41 ± 20 hrs
	Nevirapine	Viramune	25–30 hrs
NRTIs	Abacavir	Ziagen	1.5 hrs/12–26 hrs
	Didanpsine	Videx	1.5 hrs/20 hrs
	Emtricitabine	Emtriva	10 hrs/20 hrs
	Lamiduvine	Epivir	5–7 hrs
	Stavudine	Zerit	1 hr/7.5 hrs
	Tenofovir	Viread	17 hrs/60 hrs
	Zidovudine	Retrovir	1.1 hrs/7 hrs
PIs	Atazanavir	Reyataz	7 hrs
	Darunavir	Prezista	15 hrs
	Fosamprenavir	Lexiva	7.7 hrs
	Indinavir	Crixivan	1.5–2 hrs
	Lopinavir	Kaletra	5–6 hrs
	Nelfinavir	Viracept	3.5–5 hrs
	Ritonavir	Norvir	3–5 hrs
	Saquinavir	Invirase	1–2 hrs
	Tipranavir	Aptivus	6 hrs
INSTIs	Raltegravir	Isentress	9 hrs
FIs	Enfuvirtide	Fuzeon	3.8 hrs
CCR5 Antagonists	Maraviroc	Selzentry	14–18 hrs

Fusion Inhibitors and CCR5 Antagonists (FIs) interfere with the binding, fusion, and entry of an HIV virion to a human cell. There are several key proteins involved in the HIV entry process, for example, CD4, gp120, CCR5, CXCR4, and gp41. FIs have shown very promising results in clinical trials, with low incidences of relatively mild side effects. However, FIs are large molecules that must be given through injection or infusion, which limits their usefulness. The CCR5 coreceptor antagonists inhibit fusion of HIV with the host cell by blocking the interaction between the gp-120 viral glycoprotein and the CCR5 chemokine receptor [217]. The adverse events are abdominal pain, cough, dizziness, fever, and upper respiratory tract infection.

Nucleoside Reverse Transcriptase Inhibitors (NRTIs) mimic natural nucleosides, which are introduced into the DNA copy of the HIV RNA during the reverse transcription event of infection. However, the NRTIs are nonfunctional, and their inclusion terminates the formation of the DNA copy. The side effects associated with NRTIs seem to be related to mitochondrial toxicity; these are lactic acidosis, neuropathy, pancreatitis, fatigue, nausea, vomiting, and diarrhea.

Nonnucleoside Reverse Transcriptase Inhibitors (NNRTIs) also block the creation of a DNA copy of the HIV RNA but work by binding directly to key sites on the reverse transcriptase molecule blocking its action. NNRTIs do not work well on their own but in conjunction with NRTIs can increase the effectiveness of viral suppression. Side effects may include hepatoxicity, rash, dizziness, and sleepiness.

Integrase strand transfer inhibitors (INSTIs) are a class of antiretroviral drugs designed to block the action of integrase, a viral enzyme that inserts the viral genome into the DNA of the host cell. Side effects may include nausea, headache, diarrhea, and fever.

Protease inhibitors (PIs) target the viral enzyme protease that cuts the polyproteins into their respective components. With this step of viral replication blocked, the infected cell produces viral particles unable to infect cells. Available PIs can cause lipodystrophy, a severe redistribution of body fat that can drastically change the patients appearance. Other side effects include gastrointestinal disorders, nephrolithiasis, dry skin, severe diarrhea, and hepatoxicity.

7.2 GUIDELINES FOR HAART TREATMENT

The use of HAART for suppression of measurable levels of virus in the body has greatly contributed to restoration and preservation of the immune system in HIV patients. In the 1990s the dogma for HIV therapy was "Hit HIV early and hard" [218]. However, the treatment of HIV is complicated by the existence of tissue compartments and cellular reservoirs. Long-term reservoirs can survive for many years and archive many quasispecies of virus that can reemerge and propagate after withdrawal of HAART. Moreover, the virus in the central nervous system and in semen evolves independently of virus found in blood cells [219,220]. Thus, the initial enthusiasm for initiating therapy early was tempered by the recognition that standard antiretroviral therapy would probably not lead to eradication, and therefore HAART would need to be sustained indefinitely. This could be difficult for many patients due to adverse health events, metabolic complications, adherence, and costs. These short- and long-term problems associated with HAART have led to proposals for alternative treatment strategies for controlling HIV infection.

Accumulating data revealed that immune reconstitution was achievable even in those individuals with very low CD4+ T cells count, and the time to diagnosis of AIDS or mortality was not different in those individuals who were treated early. Based on this consideration, the HIV treatment guidelines until 2014 recommended that therapy should be initiated in asymptomatic patients when their CD4+ T cells count drop between 200 and 350 cell/mm^3. To reduce the morbidity and mortality associated with HIV infection, the most recent update in HIV treatment guidelines [216] suggested to initiate therapy for all individuals with HIV, regardless of CD4+ T lymphocyte cell count.

The new recommendations [216] emphasize the importance of starting HAART early and continuing treatment without interruption. In principle, HAART can be started at any time, but it is highly recommended for those asymptomatic individuals with counts at 500 cell/µl or below. Regardless of CD4+ T cells count, HAART is highly recommended to patients older than 60 years old, pregnancy, chronic hepatitis B or C, and HIV-associated kidney disease.

The DHHS (Department of Health and Human Services) panel recommends initiating antiretroviral therapy in treatment patients with one of the following types of regimen: NNRTIs + 2 NRTIs, PIs + 2 NRTIs, and INSTIs + 2 NRTIs. However, the selection of a regimen should be individualized based on virologic efficacy, toxicity, pill burden, dosing frequency, drug–drug interaction potential, resistance testing results, and comorbid conditions [216].

HIV drug resistance testing should be performed to assist in the selection of active drugs when changing HAART regimens in patients with **virologic failure**, defined as the inability to sustain suppression of HIV RNA levels to less than 50 copies/ml. The optimal virologic response to treatment is maximal virologic suppression (e.g., HIV RNA level <400 copies/ml after 24 weeks, <50 copies/ml after 48 weeks). Persistent low-level viremia (e.g., HIV RNA 50–200 copies/ml) does not necessarily indicate virologic failure or a reason to change treatment [216].

Immunologic failure can be defined as a failure to achieve and maintain an adequate CD4+ T cell response despite virologic suppression. There is no consensus for when and how to treat immunologic failure. For some patients with high treatment experience, maximal virologic suppression is not possible. In this case, HAART should be continued with regimens designed to minimize toxicity, preserve CD4+ T cell counts, and avoid clinical progression. In this scenario, an expert advice is essential and should be sought.

Antiretroviral treatment failure is defined as a suboptimal response to therapy. Treatment failure is often associated with virologic failure, immunologic failure, and/or clinical progression. Many factors are associated with an increased risk of treatment failure, including starting therapy in earlier years, presence of drug-resistant virus, prior treatment failure, incomplete medication adherence, drug side effects, toxicities, suboptimal pharmacokinetics, and other unknown reasons.

Structured treatment interruptions (STIs) consist of therapy withdrawal and re-initiation according to specific criteria. STIs were motivated in part by the clinical success of a patient in Germany, who was treated soon after diagnosis of acute HIV infection [221]. Before initiation of treatment in this patient, HIV RNA levels exceeded 80,000 copies/ml on two separate occasions, suggesting that a steady state of viremia had already been reached. After viral suppression on HAART, the therapy was temporarily discontinued, which was associated with recurrence of viremia.

However, after a second discontinuation of treatment due to concurrent hepatitis A infection, viral rebound was not observed in that patient who decided to stop therapy completely and remained virologically suppressed for the next 19 months. Since the patient's immune response progressively improved despite the absence of treatment, it was hypothesized that intermittent exposure to HIV antigens may have boosted the HIV-specific immune response in this patient via autoimmunization.

In this context, note that those individuals who have been living with HIV for at least 7 to 12 years (different authors use different time spans) and have stable CD4+ T counts of 600 or more cells/mm^3 of blood and no HIV-related diseases have been named as **Long-Term Nonprogressors (LNTP)**. However, the term LTNP is a misnomer, as it must be noted that progression toward AIDS can occur even after 15 years of stable infection [222]. The aim of the STIs mentioned earlier was either or both of (i) to stimulate the immune system to react to HIV and (ii) to allow reemergence of wild-type virus and thereby reduce problems of drug resistance.

However, a number of clinical trials of STIs [223–226] have shown adverse outcomes for patients under discontinuous therapy, including serious health risks associated with treatment interruptions. For these reasons, the recent trend in HIV treatment has been solidly against STIs.

Switching between Treatments. There has been no consensus on the optimal time to change therapy to avert or compensate for virologic failure. The most aggressive approach would be to change for any repeated, detectable viremia (e.g., two consecutive HIV RNA > 50 copies/ml after suppression). Other less conservative approaches allow detectable viremia up to an arbitrary level (e.g., 1000–500 copies/ml). However, ongoing viral replication in the presence of antiretroviral drugs promotes the selection of drug resistance mutations and may limit future treatment options [227].

Antiretroviral drug sequencing can provide a therapeutic strategy to deal with virologic failure and anticipates that therapy will fail in a proportion of patients due to resistant mutations. The primary objectives of therapy sequencing are the avoidance of accumulation of mutations and selection of multidrug-resistant viruses [228,229]. Using a mathematical model, [230] hypothesized that alternating HAART regimens, even while plasma HIV RNA levels were lower than 50 copies/ml, would further reduce the likelihood of the emergence of resistance. This concept had preliminary support from a clinical trial [228] called **SWATCH** (SWitching Antiviral Therapy Combination against HIV).

In this study, 161 patients were assigned to receive regimen A (staduvine, didanosine, efavirenz), regimen B (zidovudine, lamivudine, nelfinavir), or regimen C (alternating regimens A and B every 3 months for 12 months). Regimens A and B had the same performance, with only 20% failure rate at the end of 48-week observation. The alternating regimen outperformed both regimens A and B with only three failure events. In addition, virologic failure was noted in regimens A and B, whereas in regimen C, no resistance was documented [230]. The clinical outcome suggested that proactive switching and alternation of antiretroviral regimens with drugs that have different resistance profiles might extend the overall long-term effectiveness.

7.3 INCLUDING HAART IN MATHEMATICAL MODELS

Antiretrovirals can interfere the HIV cycle in six mechanistic classes. The most extensive study for combination regimens provides durable viral suppression and generally consists of two NRTIs plus one NNRTI or a PI [216]. Here, the previous model (6.12) is extended to include the effect of antiretrovirals:

$$
\begin{aligned}
\dot{T} &= s_T - (1 - \eta_{RT}) k_T T V - \delta_T T, \\
\dot{T^*} &= (1 - \eta_{RT}) k_T T V - \delta_{T^*} T^*, \\
\dot{M} &= s_M - (1 - f \eta_{RT}) k_M M V - \delta_M M, \\
\dot{M^*} &= (1 - f \eta_{RT}) k_M M V - \delta_{M^*} M^*, \\
\dot{V} &= (1 - \eta_{PI}) p_T T^* + (1 - f \eta_{PI}) p_M M^* - \delta_V V.
\end{aligned}
\tag{7.1}
$$

On one hand, the reverse transcriptase inhibitors (RTIs) may block infection and hence reduce the infection rate of CD4+ T cells (k_T) and macrophages (k_M). This can be represented including the term $(1 - \eta_{RT})$ into the cell infection rates. RTIs like other drugs are not perfect, and thus η_{RT} is the "effectiveness" of the reverse transcriptase inhibitors [202]. An inhibitor is efficient in the range $0 \leq a_{RT} \leq \eta_{RT} \leq b_{RT} \leq 1$, where a_{RT} and b_{RT} represent minimal and maximal drug efficacies.

On the other hand, PIs inhibit the protease of HIV, resulting in a decrease of the viral replication, which is represented including $(1 - \eta_{PI})$ in the viral replication rate of infected cells, that is, the parameters p_T and p_M. Because macrophages are long-lived cells [174], inhibitors are more effective in CD4+ T cells than in macrophages; this is contemplated using $f \in [0, 1]$.

The primary goal of antiretroviral therapy is to reduce HIV-associated morbidity and mortality. Over the last 20 years, several changes have been made to the recommendations on when to start therapy. The standard procedure for the panel of antiretroviral guidelines for adults and adolescents with HIV in USA [216] is to only make recommendations in agreement with two-thirds of the panel members. **When to start** recommendations were very debatable [231]. There is a general consensus to initiate therapy for all patients regardless of CD4+ T lymphocyte cell count. For patients with a history of an AIDS-defining illness or when CD4+ T counts are less than 350 cells/mm^3, therapy is urgently recommended [216].

Using model (7.1), simulations in Fig. 7.1 were developed assuming that HAART treatment is initiated approximately around the third year after infection. Fig. 7.1 reveals how CD4+ T cells experience a rapid depletion in counts. When treatment is introduced, CD4+ T cells counts can recover and maintain normal counts. One of the most important goals of therapy is to achieve **maximal virologic suppression**, that is, HIV RNA level less than 400 copies/ml after 24 weeks and less than 50 copies/ml after 48 weeks.

Infected cells play an important role in HIV infection. On one hand, infected CD4+ T cells show a rapid depletion when HAART is introduced, and this is maintained at a very low level for several years; see Fig. 7.1. On the other hand, infected macrophages have a very slow depletion. Therefore they can still contribute to the viral load population for a long period of time. This is consistent with many works on the area, which propose that macrophages are responsible for the second phase in the decay of plasma virus level [202].

Remark 7.1. Under the assumption of perfect effectiveness in the treatment, CD4+ T cells and macrophages dynamics can become uncoupled from viral dynamic equation. Therefore the equations can be solved for the infected cells:

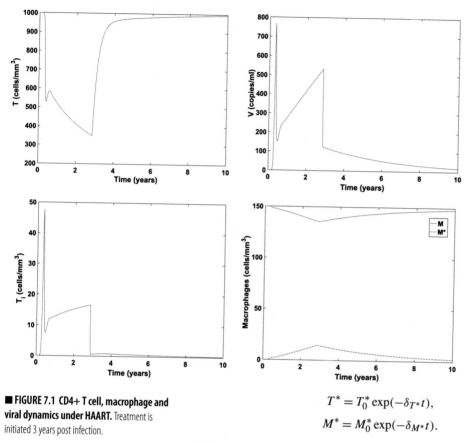

FIGURE 7.1 CD4+ T cell, macrophage and viral dynamics under HAART. Treatment is initiated 3 years post infection.

$$T^* = T_0^* \exp(-\delta_{T^*} t),$$

$$M^* = M_0^* \exp(-\delta_{M^*} t).$$

Using these equations, it is easy to see that for this model, the longer the delay to initiate HAART, the more new infections of cells and reservoirs will take place, and consequently a longer period is necessary to clear the virus.

7.4 BASIC VIRAL MUTATION TREATMENT MODELS

Numerical results show that the mathematical model (6.12) is able to represent the complete trajectory of HIV infection. Furthermore, the inclusion of treatment effects expose the fast recover in CD4+ T cells and reduction of viral load. According to simulation results, HIV infection progression can be delayed for an undetermined period of time.

Notwithstanding, HIV may mutate, and this is problematic since new genetic variants can rise to drug resistance either a single drug or a drug combination is given. Thus it is necessary to consider different genetic strains in mathematical models. For simplicity, a simple mutation model is proposed to allow control analysis and optimization of treatment switching. Based on the previous model (7.1), the following assumptions are considered.

Assumption 7.1. Constant macrophage and CD4+ T cell counts. The main nonlinearities in the general model are bilinear, and all involve either the macrophage or healthy CD4+ T cell count. Under normal treatment circumstances (i.e., after the initial infection stage and until full progression to a dominant highly resistant mutant) typical simulations and/or clinical data suggest that the macrophage and CD4+ T cell counts are approximately constant [202]. This assumption allows us to simplify the dynamics to being essentially linear.

Assumption 7.2. Scalar dynamics for each mutant. A more extensive model for HIV dynamics includes a set of states for each possible genotype such as viral strain i, CD4+ T cells infected by variant i, and macrophages infected by variant i. For simplicity, the viral strains $V_i(t)$ are the only population in the model.

Assumption 7.3. Viral clearance rate independent of treatment and mutant. In some cases, particularly in view of the earlier assumption of representing the dynamics as scalar, viral clearance rate may well depend on one or more of the treatment regimes, or on the viral genetics. For simplicity, this is considered constant for all the variants.

Assumption 7.4. Mutation rate independent of treatment and strain. It is assumed that the mutation rate between species with the same genetic distance is constant. In practice, there is some dependence of mutation rate on the replication rate, and therefore there is some relationship between genetic strain, treatment, and mutation rate.

Assumption 7.5. Deterministic model. The main interest in the next chapters is to derive control strategies with either optimal or "verifiable" performance. Therefore, to simplify the control design, mathematical models are constructed based on ODEs.

A 4 variant, 2 drug combination mathematical model. Under assumptions (7.1)–(7.5), the proposed model includes n different viral genotypes with viral strains $x_i : i = 1, \ldots, n$; N is the total number of possible drug therapies that can be administered, and $\sigma(t) \in \{1, \ldots, N\}$ is the current therapy, where σ is allowed to change at time t. The viral dynamics are represented by the equation

$$\dot{x}_i(t) = \rho_{i,\sigma(t)} x_i(t) - \delta_V x_i(t) + \sum_{j \neq i} \mu m_{i,j} x_j(t), \qquad (7.2)$$

where μ is a small parameter representing the mutation rate, the parameter δ_V is the viral clearance, and $m_{i,j} \in \{0, 1\}$ represents the genetic connections between genotypes, that is, $m_{i,j} = 1$ if and only if it is possible for genotype j to mutate into genotype i. Eq. (7.2) may be rewritten in a vector form as

$$\dot{x}(t) = \left(R_{\sigma(t)} - \delta_V I\right) x(t) + \mu M_u x(t), \tag{7.3}$$

where $M_u := [m_{ij}]$ and $R_{\sigma(t)} := diag\{\rho_{i,\sigma(t)}\}$.

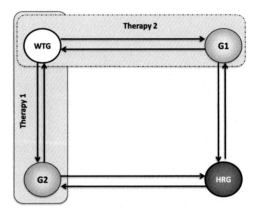

■ **FIGURE 7.2 Mutation tree for a 4 variant, 2 drug combination model**. Four genetic variants are considered. Wild-type strains can evolve in genotypes G1 and G2, which are still susceptible to therapy. Eventually, G1 and G2 can evolve into a highly resistant genotype (HRG).

As a simple motivating example, we consider a mathematical model with four genetic variants (Fig. 7.2), that is, $n = 4$, and 2 drug therapies, $N = 2$. The viral variants (also called "genotypes" or "strains") are described as follows:

- Wild Type (WT): In the absence of any therapy, this is the most prolific variant. However, it is also the variant that both drug combinations have been designed to combat, and therefore it is susceptible to both therapies.
- Genotype 1 (G1): A genotype that is resistant to therapy 1 but is susceptible to therapy 2.
- Genotype 2 (G2): A genotype that is resistant to therapy 2, but it is susceptible to therapy 1.
- Highly Resistant Genotype (HRG): A genotype with low replication rate but resistant to all drug therapies.

Viral clearance rate is fixed, $\delta_V = 0.24$ day^{-1}, which corresponds to a half life of slightly less than 3 days [202]. Typical viral mutation rates are of the order of $\mu = 10^{-4}$. A mutation graph that is symmetric and circular is assumed as in Fig. 7.2, that is, only the connections $WT \leftrightarrow G1$, $G1 \leftrightarrow HRG$, $HRG \leftrightarrow G2$, and $G2 \leftrightarrow WT$ are allowed. Other connections require double mutations and, for simplicity, are assumed to be negligible.

This leads to the mutation matrix

$$M_u = \begin{bmatrix} 0 & 1 & 1 & 0 \\ 1 & 0 & 0 & 1 \\ 1 & 0 & 0 & 1 \\ 0 & 1 & 1 & 0 \end{bmatrix}. \tag{7.4}$$

Three different scenarios for viral replication are proposed in Table 7.2. The first scenario, the most ideal case, describes a complete symmetry between $G1$ and $G2$ in the sense that therapy 1 inhibits $G2$ with the same intensity that therapy 2 inhibits $G1$. In practice, a small difference in relative replication ability is expected. Furthermore, a more detailed model also includes asymmetry in the genetic tree, which usually has a much more complex structure than a simple cycle. The second scenario shows an asymmetry for replication rates in $G1$ and $G2$ although both therapies induce the same replication in the WTP and HRG.

The more realistic case is where all genotypes experience different dynamics to the new treatment, which is represented in Scenario 3. The numerical values in Table 7.2 are of course idealized; however, the general principles are as follows:

- Genetic distance from wild type reduces fitness. In the absence of effective drug treatments, we may expect that fitness, replication rate, decreases with genetic distance from the wild type, which is expected to be the most fit.
- Therapy is at best 90% effective. In the absence of drugs, from typical data the overall viral replication rate (with high constant CD4+ T cell count) of approximately $\rho = 0.5$ day^{-1} is expected.

Table 7.2 Replication rates for viral variants and therapy combinations

Scenario	Therapy	WT (x_1)	G1 (x_2)	G2 (x_3)	HRG (x_4)
1	1	$\rho_{1,1} = 0.05$	$\rho_{2,1} = 0.27$	$\rho_{3,1} = 0.05$	$\rho_{4,1} = 0.27$
	2	$\rho_{2,1} = 0.05$	$\rho_{2,2} = 0.05$	$\rho_{3,2} = 0.27$	$\rho_{4,2} = 0.27$
2	1	$\rho_{1,1} = 0.05$	$\rho_{2,1} = 0.28$	$\rho_{3,1} = 0.01$	$\rho_{4,1} = 0.27$
	2	$\rho_{2,1} = 0.05$	$\rho_{2,2} = 0.20$	$\rho_{3,2} = 0.25$	$\rho_{4,2} = 0.27$
3	1	$\rho_{1,1} = 0.05$	$\rho_{2,1} = 0.26$	$\rho_{3,1} = 0.01$	$\rho_{4,1} = 0.29$
	2	$\rho_{2,1} = 0.01$	$\rho_{2,2} = 0.15$	$\rho_{3,2} = 0.25$	$\rho_{4,2} = 0.27$

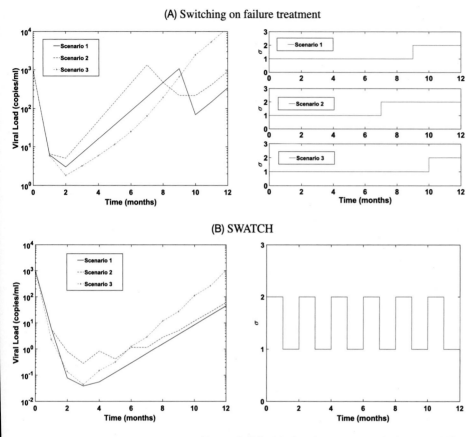

FIGURE 7.3 Switching on failure vs. SWATCH for model (7.2). Viral load is presented using different simulation scenarios.

Remark 7.2. Notice that system (7.2) is not stabilizable under the switching action. The biological reason is because the highly resistant genotype escapes the effects of treatments and immune system.

Treatments Strategies using the Basic Viral Mutation Model. "When does the patient need to change therapy?" has been a question of discussion in [216]. There are very clear factors when the patient needs to change therapy: drug side effects, prior treatment failure, comorbidities, lower CD4+ T cell counts, presence of drug-resistant virus, or other reasons. Nonetheless, there is no consensus on the optimal time to change therapy for virologic failure. Guidelines for the use of antiretroviral agents [216] recommend change for any detectable viremia up to an arbitrary level (e.g., 1000 copies/ml). Antiretroviral sequencing, called SWATCH, was proposed to deal with virologic failure and anticipates therapy failure in a proportion of patients [228]. In this clinical trial, patients were alternating between two treatments every 3 months for one-year period.

7.11 SIMULATING THE MUTATION MODEL (7.2) **FOR SCENARIO 1**

```
clear all; close all; clc
%% Set up system parameters.
mu=1e-04;                                % Mutation rate
M=[0 1 1 0;1 0 0 1;1 0 0 1;0 1 1 0];     % Mutation pattern
death_rate=0.24;                         % Viral clearance
Vinit=1000;                              % Initial condition
%% Time Simulation
dTsim=28;                                % Time step for simulations
Tmax=336;                                % Max simulation time
t=0:dTsim:Tmax; t=t';                    % Time vector
%% Reproduction rates: Scenario 1
rho=[0.05,0.05;0.27,0.05;0.05, 0.27;0.27,0.27];
R1=diag(rho(:,1)); R2=diag(rho(:,2));
%% State matrices
A1=R1-death_rate*eye(size(R1))+mu*M;
A2=R2-death_rate*eye(size(R2))+mu*M;
%% Initial Conditions
x11(1)=Vinit; x21(1)=x11(1)*mu; x31(1)=x11(1)*mu;
x41(1)=x21(1)*mu+x31(1)*mu;
z=[x11(1);x21(1);x31(1);x41(1)];
[s,g]=min([x21(1) x31(1)]);
c1(1)=g;
if g==1 At=A1; else At=A2; end

%% Main cycle for Virologic Failure
for k=1:length(t)-1;
    z=expm(At*dTsim)*z;
    x11(k+1)=z(1);x21(k+1)=z(2);x31(k+1)=z(3);x41(k+1)=z(4);
    %% Virological Failure
    if sum(z) > 1000
        if c1(1)==1 g=2; end
        if c1(1)==2 g=1; end
    end
    %% SWATCH Treatment
    % if g==1 g=2; else g=1; end
    c1(k+1)=g;
    if g==1 At=A1; else At=A2;end
end

%% Plotting
figure(1); clf;
subplot(2,1,1);
semilogy(t/28,x11,t/28,x21,t/28,x31,t/28,x41,t/28,x11+x21+x31+x41);
legend('G1','G2','G3','G4','viral load');
xlabel('Time (months)'); ylabel('copies/ml');subplot(2,1,2);
stairs(t/28,c1);
ylabel('\sigma'); xlabel('Time (months)');axis([0,Tmax/28,0,3]);
```

Note: To simulate the SWATCH strategy, the user needs to comment with % the virological failure treatment command lines and to uncomment the line after %% SWATCH treatment.

Encouraged by the optimal time to change therapy, model (7.2) is used to test these two therapeutic approaches. For numerical analysis (Matlab code in Box 7.11), it is assumed that the infected subject has a complete medical history, physical examination, and laboratory evaluation once a month for one-year period. Fig. 7.3 reveals how the total viral load initially drops rapidly for the three different scenarios. However, the appearance of resistant genotypes will drive a virologic failure (viral load > 1000 copies/ml) after 8 months, and then a new therapy is needed. Scenarios 1 and 2 exhibit a second drop in viral population, not as pronounce as it was for therapy 1. For scenario 3, the viral load explosion is almost not affected by the new therapy, which is because the HRG is directing the dynamics of the system. Using the SWATCH approach, a lower concentration in the total viral load over the year is shown in Fig. 7.3.

7.5 MODEL WITH RESERVOIRS THAT REPLICATE VIRUS FREQUENTLY

Previous simulation results suggested virologic failure before a year of treatment. This is not consistent with clinical evidence; for instance, the work [232] observed that the median time to failure was 68.4 months for patients with persistence low-viremia (51–1000 copies/ml for at least 3 months) and more than 72 months for patients without persistence low-viremia (PLV). In fact, PLV is associated with virological failure [232], that is, patients with PLV>400 copies/ml and a history of HAART experience are more likely to experience virological failure; moreover, they recommend that patients with PLV should be considered for treatment optimization and intervention studies. Accordingly, it is important to study models with more species to better match clinical observations of disease time scales. In fact, Perelson et al. [202] showed that after the first rapid phase of decay during the initial 1–2 weeks of antiretroviral treatment, plasma virus declined at a considerably slower rate. This second phase of viral decay was attributed to the turnover of a longer-lived virus reservoir. Hence the two-target cell models are more accurate than one-target cell models [202].

Now we relax Assumption 7.3 and consider other species for a more realistic model. However, the design of switching strategies for the nonlinear model (6.12) can be very demanding. For simplicity, we employ Assumption 7.1 to obtain a linear switched system of the form

$$\dot{T}_i^* = k_{T,\sigma} T V_i - \delta_{T^*} T_i^* + \sum_{j=1}^{n} \mu m_{i,j} V_j T,$$

$$\dot{M}_i^* = k_{M,\sigma} M V_i - \delta_{M^*} M_i^* + \sum_{j=1}^{n} \mu m_{i,j} V_j M, \qquad (7.5)$$

$$\dot{V}_i = p_{T,\sigma} T_i^* + p_{M,\sigma} M_i^* - \delta_V V_i,$$

where T and M are treated as constant. Here macrophages are considered as reservoirs that frequently replicate virus. The infection rate is expressed as $k_{T,\sigma}$ for CD4+ T cells and $k_{M,\sigma}$ for macrophages. Virus replicates from infected activated CD4+ T cells and infected macrophages, which is represented by $p_{T,\sigma}$ and $p_{M,\sigma}$, respectively. These parameters depend on the fitness of the genotype and therapy. The mutation rate is expressed by μ, and $m_{i,j} \in \{0, 1\}$ represents the genetic connections between genotypes. Death and clearance rates are δ_{T^*}, δ_{M^*}, and δ_V respectively. For simulation purposes, we use the parameters shown in Table 7.3.

Table 7.3 Parameters values for (7.5)

Parameter	Value	Value taken from:
k_T	3.4714×10^{-5}	Adjusted
k_M	4.533×10^{-7}	Adjusted
p_T	44	Adjusted
p_M	44	Adjusted
δ_{T^*}	0.4	[180]
δ_{M^*}	0.001	[180]
δ_V	2.4	[180]

System (7.5) can be rewritten as

$$\dot{x} = \begin{bmatrix} \Lambda_{1,\sigma} & 0 & \cdots & 0 \\ 0 & \Lambda_{2,\sigma} & \cdots & 0 \\ \vdots & & \ddots & \vdots \\ 0 & 0 & \cdots & \Lambda_{n,\sigma} \end{bmatrix} x + \mu M_u x, \qquad (7.6)$$

where $x' = [T_1^*, M_1^*, V_1, \ldots, T_n^*, M_n^*, V_n]$, $\Lambda_{j,\sigma}$ is given by

$$\Lambda_{j,\sigma} = \begin{bmatrix} -\delta_{T^*} & 0 & k_{T,\sigma} T \\ 0 & -\delta_{M^*} & k_{M,\sigma} M \\ p_{T,\sigma} & p_{M,\sigma} & -\delta_V \end{bmatrix},$$

and the mutation matrix is

$$M_u = \begin{bmatrix} m_{1,1} \begin{bmatrix} 0 & 0 & T \\ 0 & 0 & M \\ 0 & 0 & 0 \end{bmatrix} & \cdots & m_{1,j} \begin{bmatrix} 0 & 0 & T \\ 0 & 0 & M \\ 0 & 0 & 0 \end{bmatrix} \\ \vdots & \ddots & \vdots \\ m_{i,1} \begin{bmatrix} 0 & 0 & T \\ 0 & 0 & M \\ 0 & 0 & 0 \end{bmatrix} & \cdots & m_{i,j} \begin{bmatrix} 0 & 0 & T \\ 0 & 0 & M \\ 0 & 0 & 0 \end{bmatrix} \end{bmatrix}.$$

A 9 variant, 2 drug combination model. To accommodate a more complicated scenario 9, we assume genetic variants, that is, $n = 9$, and two possible cART regimens, $N = 2$. The viral variants (i) are organized in a square grid as shown in Fig. 7.4. The wild-type genotype g_1 is the most prolific variant in the absence of any drugs. However, it is also the variant that all drug combinations have been designed to combat, which is susceptible to all therapies. After several mutations, the highly resistant g_9 appears as a genotype with low replication rate but resistant to all drug therapies.

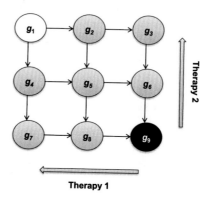

FIGURE 7.4 Mutation tree for a 9 variants, 2 drug combination model. Treatment efficacy is considered with a linear increasing factor in the direction of the arrows.

In the absence of treatment, it is considered that mutation reduces the fitness of the genotype. Thus, linear decreasing factors f_i are used to represent the fitness of genotype i. Drugs effects can be seen in Fig. 7.4, where the arrows indicate the efficiency of the drug. Moreover, based on clinical evidence [174], inhibitors are more effective in CD4+ T cells than in macrophages, which is modeled using $\eta_{\sigma,i}^T > \eta_{\sigma,i}^M$ and $\theta_{\sigma,i}^T > \theta_{\sigma,i}^M$. For simulation purposes, the linear decreasing factors on the same axis in Fig. 7.4 are fixed as $f_i = [1, 0.95, 0.95]$, and the treatment efficiencies are as follows: $\eta_{\sigma,1}^T = \theta_{\sigma,1}^T = [0.2, 0.9, 1]$, $\eta_{\sigma,2}^T = \theta_{\sigma,2}^T = [0.2, 0.5, 1]$, $\eta_{\sigma,1}^M = \theta_{\sigma,1}^M = [0.25, 0.5, 1]$, and $\eta_{\sigma,2}^M = \theta_{\sigma,2}^M = [0.1, 0.8, 1]$.

Treatments Strategies using the Macrophage Mutation Model. To evaluate extended time scales in model (7.5), longer treatment periods are assumed, and the patient has full clinical examination, once every three months as stated in [228]. Fig. 7.5 exhibits the first virologic failure after 6 years of treatment, which is consistent with clinical observations [232].

7.12 SIMULATING THE MUTATION MODEL (7.5)

```
clear all; close all; clc
%% State matrices
[A1,C1]=build_linear_model(1); [A2,C2]=build_linear_model(2);
%% Initial State variable
nx=length(C1);
Vinit=600; x0=zeros(nx,1); x0(1)=150; x0(2)=50; x0(3)=Vinit;
%% Simulation time
```

```
dTsim=30;                % Time step for simulations
dTc=30;                  % Time step for decisions
Tmax=12*30*16;           % Max simulation time
t=0:dTsim:Tmax; t=t';    % Time vector
q=C1';                   % Cost vector
N=Tmax/dTc;
%% Initialise data variables
nt=length(t); xt=zeros(nt,nx);w=zeros(nt,1);c=w;xt(1,:)=x0';z=xt(1,:)';
Phi1=expm(A1*dTsim);
Phi2=expm(A2*dTsim);
failed=0;
c(1)=1; g=1;
t_lastc=0; % Time at which the last control was computed
if g==1 Phi=Phi1; else Phi=Phi2; end
%% Simulate Viral Escape
for k=1:length(t)-1
    z=Phi*z;
    xt(k+1,:)=z';
    if (t(k+1)-t_lastc) >= dTc % time to compute a new decision
        t_lastc=t(k+1);
        [g,failed]=vir_escape(q'*z,failed,1000);
        if g==1 Phi=Phi1; else Phi=Phi2; end
    end
    c(k+1)=g;
end
%% Record data for display
C_Ti=zeros(1,27); for i=0:8; C_Ti(3*i+1)=1; end % Infected Cells
C_L=zeros(1,27); for i=0:8; C_L(3*i+2)=1; end   % Latent Cells
v_vir=xt*q; c_vir=c;
Ti=xt*C_Ti'; TL=xt*C_L';
c_vir=c;       % Switched Rule
Result_escape= v_vir(length(v_vir));
%% Figure
figure(1); clf;
subplot(3,1,1);
semilogy(t/(30*12),v_vir); ylabel('copies/ml');
subplot(3,1,2);
semilogy(t/(30*12),Ti,t/(30*12),TL );
legend('T_i','T_L'); ylabel('cells/mm^3');
subplot(3,1,3);
stairs(t/(30*12),c_vir); ylabel('\sigma(k)');
axis([0,Tmax/(30*12),0,3]); xlabel('Time (years)');
%% Function to switch when virologic failure
function [sig,failed]=vir_escape(vtotal,failed, thresh)
    if (vtotal > thresh) failed=2; end;
    if failed sig = 2; else sig = 1; end
end
function [A,C]=build_linear_model(sigma)
% 'sigma' = 1 or 2 denotes the drug combination used.
mu=3e-05;
%% Load some model parameters used in the calculations.
fitness1=[1;0.95;0.95]; fitness2=[1;0.95;0.95]; % Fitness
% Drug efficiencies for each regimen
infection_eff1=[0.2;0.9;1]; production_eff1=[0.25;0.5;1];
infection_eff2=[0.2;0.5;1]; production_eff2=[0.1;0.8;1];
%% Initializing
A=zeros(27,27); C=zeros(1,27);
for i=0:8; C(3*i+3)=1; end
def_params.KT=3.4714e-5;  def_params.dT=0.4;    def_params.PT=44;
def_params.KM=4.533e-7;   def_params.dM=0.001;  def_params.PM=44;
def_params.T=600;         def_params.M=140;     def_params.dV=2.4;

%% Build individual SS matrices
for i1=1:3
    for i2=1:3
        i=i1+3*i2-3; %Model subindex, i=1,2..9
        params=def_params;
        if sigma ==1
            params.KT = params.KT * infection_eff1(i1);
            params.KM = params.KM * infection_eff1(i1);
            params.PT = params.PT * fitness1(i1)*fitness2(i2)*production_eff1(i1);
            params.PM = params.PM * fitness1(i1)*fitness2(i2)*production_eff1(i1);
        else
            params.KT = params.KT * infection_eff2(i2);
            params.KM = params.KM * infection_eff2(i2);
```

```
                params.PT = params.PT * fitness1(i1)*fitness2(i2)*production_eff2(i2);
                params.PM = params.PM * fitness1(i1)*fitness2(i2)*production_eff2(i2);
            end
            [Ai,Bi,Ci]=mutant_model(params);
            CC(:,:,i)=Ci;
            A(i*3-2:i*3,i*3-2:i*3)=Ai;
        end
    end
    %% Define mutation matrix. First define edges in the graph
    edges=[    1,2;    1,4;  2,3;    2,5;    3,6;
               4,5;    4,7;  5,6;    5,8;    6,9;
               7,8;    8,9];
    for i=1:length(edges)
        i1=edges(i,1); i2=edges(i,2);
        N1=i1*3-2:i1*3; N2=i2*3-2:i2*3;
        A(N1,N2)=mu*CC(:,:,i2); A(N2,N1)=mu*CC(:,:,i1);
    end
    End

    function [A,B,C]=mutant_model(params)
    B=[1;0;0];
    A=[ -params.dT,        0,           params.KT*params.T;
        0,             -params.dM,      params.KM*params.M;
        params.PT,     params.PM,       -params.dV,];
    C=A; C(1:3,1:2)=0; C(3,3)=0;
    end
```

The numerical results in Fig. 7.5, Matlab code in Box 7.12, highlight two important observations. One is that infected CD4+ T cells show very similar dynamics to the virus, which is because viruses rapidly infect new CD4+ T cells. Secondly, infected macrophages present a more robust behavior against treatments, which reveals a hint that macrophages, or any other reservoirs that replicate virus frequently, can persist even during long periods of HAART treatment. Therefore maintaining therapy in patients is important to avoid the formation of latent reservoirs, which will continue to experience viral replication. The simulation results also suggested that both therapies will last for approximately 14 years before a virological failure can occur.

In contrast to virologic failure treatment, the SWATCH approach can maintain the total viral load under virologic failure levels for roughly 16 years; see Fig. 7.5. This gives an extension of nearly 2 years compared with virological failure treatment. These simulation results show that the recycling therapies can extend the use of HAART treatments. Similar evidence was found in [228], comparing to a group of patients with virologic failure treatment, significantly more patients in the alternating therapy group (SWATCH) had plasma HIV-1 RNA levels less than 400 copies/ml.

7.6 MUTATION MODEL WITH LATENT CD4+ T RESERVOIRS

Simulations reveal that model (7.5) exhibited time scales consistent with clinical observations. Here it is explored whether another mechanism can match clinical evidence. Using the latently infected CD4+ T cells, another

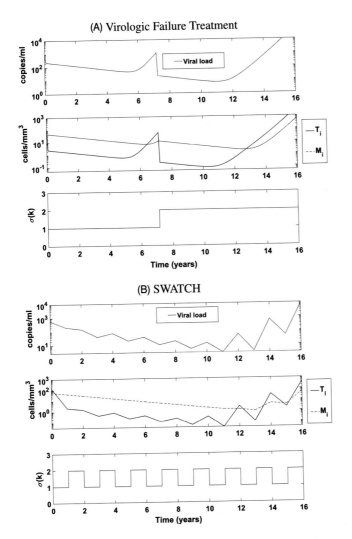

■ **FIGURE 7.5 Virologic failure Treatment vs. SWATCH for a 9 genotypes model.** Viral dynamics and infected cells are presented.

type of HIV reservoirs, may have an important role in the late HIV infection stage, the model proposed in [202] is adopted. Using Assumption 7.1, we derive the following model:

$$\dot{T}_i^* = \psi k_{T,\sigma} T V_i + a_L L_i - \delta_{T^*} T_i^* + \sum_{j=1}^{n} \mu m_{i,j} V_j T,$$

$$\dot{L}_i = (1 - \psi) k_{T,\sigma} T V_i - a_L L_i - \delta_L L_i, \qquad (7.7)$$

$$\dot{V}_i = p_{T,\sigma} T_i^* - \delta_V V_i,$$

where L_i, represents the latently infected CD4+ T cells. The infection rate is expressed by $k_{T,\sigma}$. This parameter depends on the genotype and the therapy used. Once CD4+ T cells are infected, a proportion of cells ψ passes into the infected cells population, whereas a proportion $(1 - \psi)$ passes into the latently infected cell population. These latently infected cells may be activated later and start the virus replication, which is represented by the term a_L. Viral replication is achieved in infected activated CD4+ T cells, which is represented by $p_{T,\sigma}$ depending on the fitness of the genotype and the therapy. The mutation rate is represented by μ, the death rates are represented by δ_{T^*}, δ_L, and δ_V respectively, and $m_{i,j} \in \{0, 1\}$ represents the genetic connections between genotypes. The parameters of Table 7.4 are used for simulations.

Table 7.4 Parameters values for (7.7)

Parameter	Value	Value taken from:
k_T	3.8×10^{-3}	Adjusted
p_T	6.45×10^{-1}	Adjusted
δ_{T^*}	0.4	[180]
δ_V	2.4	[180]
δ_L	0.005	[180]
a_L	3×10^{-4}	[180]
ψ	0.97	[180]

Assumption 7.1 allows us to simplify the dynamics to a linear system. Therefore system (7.7) can be rewritten as (7.6), where $x' = [T_1^*, L_1, V_1, \ldots, T_n^*, L_n, V_n]$, $\Lambda_{j,\sigma}$ is given by

$$\Lambda_{j,\sigma} = \begin{bmatrix} -\delta_{T^*} & a_L & \psi k_{T,\sigma} T \\ 0 & -(a_L + \delta_L) & (1 - \psi)k_{T,\sigma} T \\ p_{T,\sigma} & 0 & -(c_T T + \delta_V) \end{bmatrix},$$

and the mutation matrix is

$$M_u = \begin{bmatrix} m_{1,1} \begin{bmatrix} 0 & 0 & T \\ 0 & 0 & 0 \\ 0 & 0 & 0 \end{bmatrix} & \cdots & m_{1,j} \begin{bmatrix} 0 & 0 & T \\ 0 & 0 & 0 \\ 0 & 0 & 0 \end{bmatrix} \\ \vdots & \ddots & \vdots \\ m_{i,1} \begin{bmatrix} 0 & 0 & T \\ 0 & 0 & 0 \\ 0 & 0 & 0 \end{bmatrix} & \cdots & m_{i,j} \begin{bmatrix} 0 & 0 & T \\ 0 & 0 & 0 \\ 0 & 0 & 0 \end{bmatrix} \end{bmatrix}.$$

A 64 variant, 3 drug combination mathematical model. A mutation model with 64 genetic variants is proposed, that is, $n = 64$, and three possible drug therapies ($N = 3$). The viral variants are organized in a three-dimensional lattice as shown in Fig. 7.6. This lattice is based on the simplifying assumption that multiple independent mutations are required to achieve resistance to all therapies.

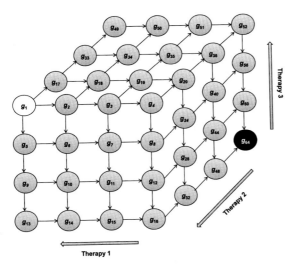

■ **FIGURE 7.6 Three-lattice mutation tree.** Mutation tree for a 64 genetic variants and 3-drug combination model. Treatment efficacy is considered with a linear increasing factor in the direction of the arrows.

As with the previous models, the wild-type genotype g_1 would be the most prolific variant in the absence of any drugs; however, it is also a variant that is susceptible to all therapies. After several mutations, the highly resistant genotype g_{64} appears as a genotype with low replication rate but resistant to all drug therapies.

In a similar vein to previous models, therapies are composed of reverse transcriptase inhibitors and protease inhibitors, that is,

$$k_{T,\sigma} = k_T f_j \eta_{\sigma,j}, \tag{7.8}$$

$$p_{T,\sigma} = p_T f_j \theta_{\sigma,j}, \tag{7.9}$$

where $\eta_{\sigma,j}$ represents the infection efficiency for genotype j under treatment σ, and $\theta_{\sigma,j}$ expresses the replication efficiency for the genotype j under treatment σ. Similarly, the absence of treatment mutation reduces the fitness of the genotype considered, and thus linear decreasing factors for f_j are used to represent the fitness of genotype j.

The drug efficiency gradients are illustrated in Fig. 7.6, where the arrows indicate the efficiency of the drug. For instance, the genotypes $g_1, g_5, \ldots, g_{61}$

are all on one face of the lattice and are fully susceptible to therapy 1. The opposite face $g_4, g_8, \ldots, g_{64}$ describes all genotypes highly resistant to therapy 1. For simulation purposes, linear decreasing factors $f_i = [1, 0.98, 0.96, 0.94]$ are considered on the same axis in the three lattices as shown in Fig. 7.6, and the treatment efficiencies are as follows: $\eta^T_{\sigma,1} = \theta^T_{\sigma,1} = [0.5, 0.6, 0.75, 1]$, $\eta^T_{\sigma,2} = \theta^T_{\sigma,2} = [0.2, 0.5, 0.7, 1]$, and $\eta^T_{\sigma,3} = \theta^T_{\sigma,3} = [0.3, 0.6, 0.85, 1]$.

Treatments Strategies using the model with Latent CD4+ T Reservoirs.
A more complex mutation tree model to evaluate the switch on failure treatment and the SWATCH therapy is employed. Using 64 genotypes and three therapies, longer treatment periods are considered assuming that the patient has full clinical examination, once every three months.

Fig. 7.7 shows how the first virologic failure occurs around the first year of treatment. Active infected CD4+ T cells have a very similar dynamics to the viral load, which is because in activated CD4+ T cells, viral replication is rapid and efficient. On the other hand, latently infected cells show a more robust behavior against treatments, where patients under HAART provided the first real evidence that HIV-1 can persist in a latent form, and then it can be rescued from cells after their activation.

In contrast to virological failure treatment, the SWATCH approach (switching between regimens every three months) can maintain the total viral load under detectable level for approximately 10 years, and the virologic failure appears after 14 years. Qualitatively, similar evidence was found in [228], where significantly more patients in the alternating therapy group than in the virologic failure treatment group had plasma HIV RNA levels less than 400 copies/ml while receiving treatment. It can be inferred that switching regimens affect stronger latently infected cells, and, as a result, low viral load is prolonged for a longer period than that resulted from the virologic failure treatment.

7.7 CONCLUDING REMARKS

In this chapter, we described different mathematical models to explore how to avoid or delay the appearance of highly resistant genotypes using HAART treatments. This is still an open problem of discussion in clinical circles, but using mathematical models, new insights into HIV infection can be offered.

Treatment terms were included in the mathematical model. Simulation results suggested the importance of starting HAART treatment in early stages of the infection to avoid new cell infections, especially for reservoirs that

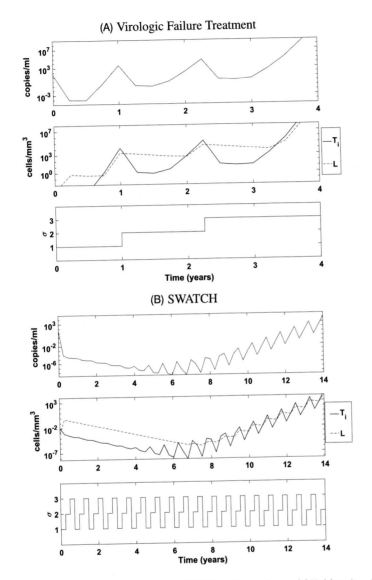

FIGURE 7.7 Virologic failure Treatment vs. SWATCH for a 64 genotypes model. Viral dynamics and infected cells are presented.

will take a long time to be cleared. For a patient under treatment, three different models were proposed to explore how the switch on regimens should be implemented. Two clinical treatments were considered, the switched on failure, which is commonly used in clinical practice, and SWATCH. Simulations exposed the importance of proactive switching to decrease viral load maximally.

The simulation results also uncovered that even with the complexity of the mutation tree presented in Fig. 7.6, time scales for virus dynamics under treatment match better in the macrophages linear model (7.5). Therefore, the long-term behavior of macrophages or any reservoir that replicate virus frequently is necessary to obtain the appropriate dynamics.

Part 3

Advanced Topics in Control Theory

Chapter 8

Optimal Therapy Scheduling

CONTENTS

8.1 OPTIMAL CONTROL BACKGROUND

Dynamic optimization has receive particular attention within the framework of control theory [233]. *The goal of optimal control* is to determine the control signals that will cause a process satisfying physical constraints and at the same time minimizing (or maximizing) some performance criterion. Let us consider the equation

$$\Sigma_F : \dot{x}(t) = F(x(t), u(t), t), \tag{8.1}$$

where $x \in R^n$, $u \in R^m$, and F is a continuously differentiable function. The initial condition is $x(t_0) = x_0$. The set $S_f \subseteq \{(x, t) \mid x \in R^n \ t \geq t_0\}$ specifies the admissible final events, so the condition $(x(t_f), t_f) \in S_f$ allows us to deal with a variety of meaningful cases: (i) the final time t_f given and the final state free, (ii) the final state given and the final time free. The set of functions that can be selected as inputs to the system coincides with U^m and the space of piecewise continuous m-vector functions. Thus the constraint $u(\cdot) \in U^m$ must be satisfied. To fully define the optimal control problem, it is necessary to consider the performance criterion for the system, which is assumed to be of the form

$$J = \int_{t_0}^{t_f} g(x(t), u(t), t)dt + h(x(t_f), t_f), \tag{8.2}$$

where g and h are continuously differentiable functions. This cost function measures the penalty that must be paid as a consequence of the dynamic

Modeling and Control of Infectious Diseases in the Host. https://doi.org/10.1016/B978-0-12-813052-0.00020-8

system trajectory. A cost function containing only a terminal penalty is called Mayer type, one with just the integral term is of the Lagrange type, whereas Bolza type represents both terms. By denoting the solution at time t as $\varphi(t; t_0, x_0, u(\cdot))$ the optimal control problem can be stated as follows.

Problem 8.1. *Optimal Control Problem.*

Determine a vector $u^o(\cdot)$ of functions defined on the interval $[t_0, t_f^o]$ so that (8.1) is satisfied and the performance index (8.2) evaluated for $x(\cdot) = \varphi((\cdot); t_0, x_0, u^o(\cdot))$ and $u(\cdot) = u^o(\cdot)$ over the interval $[t_0, t_f^o]$ takes the least possible value.

The solution for the optimal control problem is mainly based on two techniques. The Hamilton–Jacobi theory supplies sufficient conditions for the global optimality together with its most significant achievement, namely the solution of a linear quadratic (LQ) problem. The second technique, the maximum principle, exploits a first-order variational approach and provides powerful necessary conditions suited to encompass a wide class of fairly complex problems. The following optimal control background was taken from different sources [233–237].

Hamilton–Jacobi Theory

The Hamilton–Jacobi method leads to a nonlinear partial differential equation, which is computationally difficult. To establish the optimal control problem based on Hamilton–Jacobi theory, we present the following definitions.

Definition 8.1. Given the interval $[t_0, t_f^o]$ and $t_f^o \geq t_0$, $u(\cdot) \in U^m$ is an admissible control relative to (x_0, t_0) for system (8.1), the performance index (8.2), and the set S_f if

(i) $(\varphi(t_f^o; t_0, x_0, u^o(\cdot)), t_f^o) \in S_f$ and

(ii) $J(\varphi(\cdot; t_0, x_0, u^o(\cdot)), u^o(\cdot), t_f^o) \leq J(\varphi(\cdot; t_0, x_0, u(\cdot)), u(\cdot), t_f,)$.

Here $u(\cdot) \in U^m$ defined on the interval $[t_0, t_f]$ is any admissible control relative to (x_0, t_0) for system (8.1) and the set S_f.

A fundamental function in optimal control is the Hamiltonian, which is formulated from the system to be controlled and the integral part of the performance index [233].

Definition 8.2. The **Hamiltonian function** is defined as

$$H(x, u, t, \pi) := g(x, y, t) + \pi' F(x, u, t) \tag{8.3}$$

for $\pi \in R^n$.

Definition 8.3. The Hamiltonian function is said to be regular if it admits, for all x, π, and $t \geq t_0$, a unique absolute minimum $u_h^o(x, t, \pi)$, that is, if for all $u \neq u_h^o$,

$$H(x, u_h^o(x, t, \pi), t, \pi) < H(x, u, t, \pi). \tag{8.4}$$

Definition 8.4. Assuming that the Hamiltonian function is regular, the Hamilton–Jacobi equation (HJE) is given by

$$\bar{V}_t(x(t), t) + H(x(t), u(x(t), \bar{V}_x, t), \bar{V}_x, t) = 0, \tag{8.5}$$

where $\bar{V}_t = \frac{\partial \bar{V}}{\partial t}$ and $\bar{V}_x = \frac{\partial \bar{V}}{\partial x}$.

The value function $\bar{V}(t, x)$ provides, for any given state x at any given time t, the smallest possible cost among all possible trajectories starting at this event. This function is sometimes called the optimal return function or "Cost to Go" because it measures the cost of completing the trajectory. Now we can state a sufficient condition for optimality [234].

Theorem 8.1. *Let the Hamiltonian function* (8.3) *be regular, and let* $u^o \in U^m$, *defined on the interval* $[t_0, t_f^o]$, $t_f^o \geq t_0$, *be an admissible control relative to* (x_0, t_0) *such that* $(x^o(t_f^o), t_f^o) \in S_f$, *where* $x^o(\cdot) := \varphi(\cdot; t_0, x_0, u^o(\cdot))$. *Let* $\bar{V}$ *be a solution of* (8.5) *such that:*

(i) it is continuously differentiable,

(ii) $\bar{V}(z, t) = h(z, t) \; \forall (z, t) \in S_f$, *and*

(iii) $u^o(t) = u_h^o(x^o(t), t, \bar{V}_z'|_{z=x^o(t)})$.

Then $u^o(\cdot)$ *is an optimal control relative to* (x_0, t_0) *with* $J(x_0, t_0, t_f^o, x^o(\cdot), u(\cdot)) = \bar{V}(x_0, t_0)$ *[234].*

Note that if there is a function $\bar{V}(t, x)$ that satisfies the HJE, then $\bar{V}(t, x)$ is the minimum cost function [234]. In some cases, a solution is obtained by guessing a form for the minimum cost function. Unfortunately, in general, the HJE must be solved by numerical techniques.

Maximum Principle

Variational methods are used to derive necessary conditions for optimal control assuming that the admissible controls are not bounded. The principle of Pontryagin proposed by the Russian mathematician Pontryagin (1908–1988) is used for bounded control and state variables. An excellent reference for this section is [235]. Before describing the maximum principle, important concepts are defined.

Definition 8.5. **A functional** J_f is a rule of correspondence that assigns to each function f in a certain class Ψ a unique real number. Ψ is called the domain of the functional, and the set of real numbers associated with the functions in Ψ is called the range of the functional. Notice that the domain of a functional is a class of functions that is a "function of a function."

Definition 8.6. *Increment and Variation*

If f and $f + \delta f$ are functions for which the functional J_f is defined, then the **increment** of J_f, denoted by ΔJ_f, is

$$\Delta J_f = J_f(f + \delta f) - J_f(f).$$

Then $\Delta J_f(f, \delta f)$ can be written to emphasize that the increment depends on the functions f and δf; δf is called the **variation** of the function f.

Definition 8.7. *Relative Extremum*

A function f with domain D_f has a **relative extremum** at the point q^* if there is $\epsilon > 0$ such that, for all points q in D_f that satisfy $\|q - q^*\| < \epsilon$, the increment of f has the same sign [235].

Theorem 8.2. *Fundamental Theorem of Calculus of Variations*

Let f be a vector function of t in the class Ψ, and let J_f be a differentiable functional of f. Assume that the functions in Ψ are not constrained by any boundaries. Then, if f^* is an extremal, then the variation of J_f vanishes on f^*, that is,

$$\delta J(f^*, \delta f) = 0$$

for all admissible δf.

Let us now proceed with the formulation of the maximum principle theorem, which yields the solution of the fundamental problem. To formulate the theorem, in addition to the fundamental system (8.1), we consider another set of equations and auxiliary variables $\pi_0, \pi_1, ..., \pi_n$.

Definition 8.8. *Adjoint or Auxiliary System*

With reference to the Hamiltonian function (8.3) and a pair of functions $(x(\cdot), u(\cdot))$ satisfying (8.1), the system of linear differential equations

$$\dot{\pi} = -\frac{\partial H(x, u(t), t, \pi_0, \pi(t))}{\partial x}$$

is called the auxiliary system.

For fixed values of π and x, the function H becomes a function of the parameter $u \in U^m$. Let us denote the upper bound of the values of this function by $\mathcal{M}(x, \pi)$:

$$\mathcal{M}(x, \pi) = \sup_{u \in U} H(x, u, \pi).$$

If the continuous function H achieves its upper bound on U^m, then $\mathcal{M}(x, \pi)$ is the maximum value of H for fixed π and x. This can be formulated as the following theorem [235].

Theorem 8.3. *Maximum Principle Theorem*

Let $u(t)$, $t_0 \leq t \leq t_f$, be an admissible control that transfers the phase point from x_0 to x_f, and let $x(t)$ be the corresponding trajectory, so that $x(t_0) = x_0$ and $x(t_f) = x_f$. In order that $u(t)$ and $x(t)$ be time-optimal, it is necessary that there exist a nonzero continuous vector function $\pi(t) = (\pi_1, \pi_2, ..., \pi_n(t))$ corresponding to $u(t)$ and $x(t)$ such that:

1° for all $t_0 \leq t \leq t_1$, the function $H(\pi(t), x(t), u)$ of the variable $u \in U^m$ attains its maximum at the point $u = u(t)$:

$$H(\pi(t), x(t), u(t)) = \mathcal{M}(\pi(t), x(t)); \qquad (8.6)$$

2° at the terminal time t_f, the relation

$$\mathcal{M}(\pi(t_f), x(t_f)) \geq 0 \qquad (8.7)$$

is satisfied [235]. Furthermore, it turns out that if $\pi(t)$, $x(t)$, and $u(t)$ satisfy

$$\frac{\partial x}{\partial t} = \frac{\partial H}{\partial \pi},$$
$$\frac{\partial \pi}{\partial t} = -\frac{\partial H}{\partial x},$$

and condition 1°, then the time function $\mathcal{M}(\psi(t), x(t))$ is constant. Thus (8.7) may be verified at any time $t_0 \leq t \leq t_1$ and not just at t_1.

8.2 POSITIVE SWITCHED LINEAR SYSTEMS

Dynamical systems that are described by an interaction between continuous and discrete dynamics are usually called **hybrid systems** [238]. Continuous-time systems with (isolated) discrete switching events are referred as **switched systems**. A switched system may be obtained from the hybrid system neglecting the details of the discrete behavior and instead considering all possible switchings from a certain class. These systems have the important property that any nonnegative input and nonnegative initial state generates a nonnegative state trajectory and output for all future times. Common examples of positive systems include chemical processes, biology, economics, and sociology [239]. The following positive switched linear system is considered on a finite time interval:

$$\Sigma_A : \dot{x}(t) = A_{\sigma(t)}x(t), \quad x(0) = x_0, \tag{8.8}$$

where $A_{\sigma(t)}$ switches between some given finite collection of matrices $A_1, ..., A_N, t \geq 0, x(t) \in R_+^n$ is the state variable vector, $x_0 \in R_n^+$, and $\sigma(t)$ is the switching signal. This is a piecewise constant function σ that has a finite number of discontinuities, called the **switching times**, and takes a constant value on every interval between two consecutive switching times.

Theorem 8.4. *System (8.8) is positive if and only if the matrices A_i are Metzler, that is, their nondiagonal elements are nonnegative. Then, for every nonnegative initial state and every nonnegative input, its state and output are nonnegative [239].*

Definition 8.9. A **convex hull** for a set of points X in a real vector space is the minimal convex set containing X. Thus $co(X_1, X_2, \cdots, X_N)$ denotes a convex hull of matrices X_i.

Definition 8.10. A continuous vector-valued function $\Phi(t)$ defined on $[0, T]$ is a *solution in the sense of Filippov* of $\dot{x} = f(x)$ if $\Phi(t) \in K\{f(\Phi(t))\}$ for almost all t and the initial condition $\Phi(0) = x(0)$ is satisfied. Briefly, K is an average set of vector directions f that are collected near x, excluding points in a set of measure zero.

Notation. Throughout this chapter, we adopt the following notation: (a) for $x \in R^n$, $x \succ 0 (x \succeq 0)$ means that $x_i > 0 (x_i \geq 0)$ for $1 \leq i \leq n$, (b) for $A \in R^{n \times n}$, $A \succ 0 (A \succeq 0)$ means that $a_{ij} > 0 (a_{ij} \geq 0)$ for $1 \leq i, j \leq n$, and (c) for $x, y \in R^n$, $x \succ y (x \succeq y)$ means that $x - y \succeq 0 (x - y \succeq 0)$. The exponential matrix of A is represented by e^A. The symbol $\mathcal{S}_{gn}$ denotes the sign function, which takes the value 1 when its argument is positive, -1 when its argument is negative, and 0 when its argument is 0.

8.3 OPTIMAL CONTROL FOR POSITIVE SWITCHED SYSTEMS

Beginning in the late 1950s and continuing today, the issues concerning dynamic optimization have received a lot of attention within the framework of control theory [233]. The goal of optimal control goal is to determine the control signals that cause a process to satisfy the physical constraints and at the same time minimize (or maximize) some performance criterion. The solution for the optimal control problem can be developed by the Hamilton–Jacobi theory to supply sufficient conditions or by the maximum principle to derive necessary conditions. A complete background in optimal control can be found in [233–237].

The problem of determining optimal switching trajectories in hybrid systems has been widely investigated, from both theoretical and computational points of view [240–242]. For continuous-time switched systems, several works present necessary and/or sufficient conditions for a trajectory to be optimal using the minimum principle [243,244]. However, there is no general solution for the optimal control problem.

In this chapter, we study a specific class of autonomous switched systems, where the continuous control is absent, and only the switching signal must be determined [242]. The switched system may be embedded into a larger family of nonlinear systems. Sufficient conditions for optimality on a finite horizon are developed using [237]. No constraints are imposed on the switching, and the performance index contains no penalty on the switching. When dealing with switched systems, it is common to encounter sliding trajectories, that is, infinite frequency switching of $\sigma(t)$. To include sliding trajectories, we embed the switched system in the larger class described by

$$\dot{x}(t) = \sum_{i=1}^{N} u_i(t) A_i x(t) \tag{8.9}$$

with $u(t) \in \mathcal{U} := \left\{ u : u_i \geq 0 \ \forall i; \sum_{i=1}^{N} u_i = 1 \right\}$ (the unit simplex).

Definition 8.11. A **simplex** is the generalization of a tetrahedral region of space to n dimensions.

Remark 8.1. Clearly, by construction system (8.9) includes system (8.8). It may be that, for some t, $u(t)$ is not a vertex of the simplex and there is no directly equivalent $\sigma(t)$. However, note that the set of possible trajectories of (8.8) are dense in the set of trajectories generated by (8.9). Therefore, extending the concept of valid switching signals to sliding modes based on the appropriate differential inclusions, the optimal control of system (8.9) is included [245].

The cost functional to be minimized over all admissible switching sequences is given by

$$J(x_0, x, \sigma) = \int_0^{t_f} q'_{\sigma(\tau)} x(t) dt + c' x(t_f), \tag{8.10}$$

where $x(t)$ is a solution of (8.8) with switching signal $\sigma(t)$. The vectors q_i are assumed to have nonnegative entries, and c is assumed to have all positive entries. The optimal switching signal, the corresponding trajectory, and the optimal cost functional are denoted as $\sigma^o(t, x_0)$, $x^o(t)$, and $J(x_0, x^o, \sigma^o)$, respectively. The Hamiltonian function relative to system (8.8) and the cost functional (8.10) is given by

$$H(x, \sigma, \pi) = q'_\sigma x(t) + \pi' A_\sigma x(t). \tag{8.11}$$

Theorem 8.5. *Continuous-Time Optimal Control for Positive Switched Systems*

Let $\sigma^o(t, x_0) : [0, t_f] \times R^n_+ \to \mathcal{I} = \{1, \ldots, N\}$ be a switching signal relative to x_0, and let $x^o(t)$ be the corresponding trajectory. Let $\pi^o(t)$ denote a positive vector solution of the system of differential equations

$$\dot{x}^o(t) \;\; = \;\; A_{\sigma^o(t,x_0)} x^o(t), \tag{8.12}$$

$$-\dot{\pi}^o(t) \;\; = \;\; A'_{\sigma^o(t,x_0)} \pi^o(t) + q_{\sigma^o(t,x_0)}, \tag{8.13}$$

$$\sigma^o(t, x_0) \;\; = \;\; \underset{i \in \mathcal{I}}{\text{argmin}} \{\pi^{o'}(t) A_i x^o(t) + q'_i x^o(t)\} \tag{8.14}$$

with the boundary conditions $x^o(0) = x_0$ and $\pi^o(t_f) = c$. Then $\sigma^o(t, x_0)$ is an optimal switching signal relative to x_0, and the value of the optimal cost functional is

$$J(x_0, x^o, \sigma^o) = \pi'^o(0) x_0. \tag{8.15}$$

Proof. The scalar function

$$v(x, t) = \pi^o(t)' x(t) \tag{8.16}$$

is a generalized solution of the HJE

$$0 = \frac{\partial v}{\partial t}(x, t) + H\left(x(t), \sigma^o(t, x_0), \frac{\partial v}{\partial x}(x, t)'\right), \tag{8.17}$$

where

$$H(x, \sigma, \pi) = q'_\sigma x(t) + \pi'(t) A_\sigma x(t). \tag{8.18}$$

Note that the triple (x^o, π^o, σ^o) satisfies the necessary conditions of the Pontryagin principle, since

$$H(x^o, \sigma^o, \pi^o) \leq H(x, \sigma, \pi), \quad \sigma = 1, 2.$$

Moreover,

$$\frac{\partial v}{\partial x}(x, t) = \pi^o(t)', \tag{8.19}$$

$$\frac{\partial v}{\partial t}(x, t) = \dot{\pi}^o(t)' x(t), \tag{8.20}$$

so that, for almost all $t \in [0, t_f]$,

$$\dot{\pi}^o(t)' x^o(t) + q'_{\sigma^o(t, x_0)} x^o(t) + \pi^o(t)' A_{\sigma^o(t, x_0)} x^o(t) = 0. \tag{8.21}$$

Moreover, it satisfies the boundary condition

$$v(x^o(t_f), t_f) = \pi^o(t_f)' x^o(t_f) = c' x^o(t_f). \tag{8.22}$$

This completes the proof. $\qquad\qquad\qquad\square$

The HJE partial differential equation is reduced to a set of ordinary differential equations (8.13). Note that these equations are inherently nonlinear. The state equations must be integrated forward, whereas the costate equations must be integrated backward, both according to the coupling condition given by the switching rule. As a result, the problem is a two-point boundary value problem and cannot be solved using regular integration techniques. Next, we discuss how system (8.13) can be solved analytically under certain assumptions on the matrix A_σ.

8.4 OPTIMAL CONTROL TO MITIGATE HIV ESCAPE

Whereas the CD4+ T cell count is the strongest predictor of subsequent disease progression [216], the viral load is the most important indicator of response to antiretroviral therapy. Analysis of 18 clinical trials with viral load monitoring showed significant association between a decrease in plasma viremia and improved clinical outcome [246]. Therefore we can infer that both viral load and CD4+ T cells are critical. However, mostly during treatments and numerical simulations, viral load is low, and CD4+ T cell count is good (over 350 cells/mm^3) until the final viral escape appears. Moreover, the final escape of the virus is at some almost uncontrollable exponential rate, and final viral load is a surrogate for time to escape. Consequently, it is reasonable to assume that if the total viral load is small enough during a finite time of treatment, then new cell infections are less likely. Therefore there is a significant probability that the total virus load becomes zero. Observe that in a more accurate stochastic model of viral dynamics, $x_i(t)$ is the expected value of the number of viruses V_i. Hence, from Markov's inequality it can be shown that small $E[x]$ guarantees a high probability of viral extinction ($P(\sum_i v_i = 0) \geq 1 - E[\sum_i v_i] = 1 - \sum_i x_i$). It is therefore logical to propose the cost

$$J := c' x(t_f) \tag{8.23}$$

where c is the column vector with all ones, and t_f is the final time for the treatment. This cost is minimized under the action of the switching rule.

Definition 8.12. A triple $u^o(t) : [0, t_f] \times \mathcal{U}, x^o(t), \pi^o(t)$ that satisfies (for almost all t) the system of equations

$$\dot{x}^o(t) = \sum_{i=1}^N u_i^o(t) A_i x^o(t), \tag{8.24}$$

$$-\dot{\pi}^o(t) = \sum_{i=1}^N u_i^o(t) A_i' \pi^o(t), \tag{8.25}$$

$$u^o(t) \in \operatorname*{argmin}_{u \in \mathcal{U}} \{\pi^{o\prime}(t) \sum_{i=1}^N u_i A_i x^o(t)\} \tag{8.26}$$

with boundary conditions $x^o(0) = x_0$ and $\pi^o(t_f) = c$ is called a *Pontryagin solution* for the optimal control problem.

Theorem 8.6. *Assume that there exists a unique Pontryagin solution (u^o, x^o, π^o) for the optimal control defined by system (8.9) and cost (8.23). Then $u^o(t)$ is an optimal control signal relative to x_0, and the value of the optimal cost functional is $\pi^{o\prime}(0)x_0$.*

Proof. Write the Hamiltonian function

$$H(x, u, \pi) = \pi(t)' \sum_{i=1}^N u_i A_i x(t)$$

and notice that

$$\dot{\pi}(t) = -\left(\frac{\partial H}{\partial x}\right)' = -\sum_{i=1}^N u_i(t) A_i' \pi(t),$$

$$\dot{x}(t) = \left(\frac{\partial H}{\partial \pi}\right)' = \sum_{i=1}^N u_i(t) A_i x(t).$$

Moreover, the transversal conditions are satisfied, and $H(x^o, u^o, \pi^o) \leq H(x^o, u, \pi^o)$ for all $u \in \mathcal{U}$. Hence, in view of the Pontryagin principle, the triple (x^o, π^o, u^o) satisfies the necessary conditions for optimality. Optimality follows from the assumed uniqueness of the Pontryagin triple. $\square$

Remark 8.2. For almost all t, the scalar function $v(x, t) = \pi^o(t)'x$ satisfies

$$0 = \tfrac{\partial v}{\partial t}(x^o(t), t) + \min_u H\left(x^o(t), u, \tfrac{\partial v}{\partial x}(x^o(t), t)'\right)$$

with boundary condition

$$v(x^o(t_f), t_f) = \pi^o(t_f)'x^o(t_f) = c'x^o(t_f).$$

This is however not enough to guarantee that any Pontryagin solution is also optimal, even though in the literature there is no counterexample for switched linear positive systems with positive linear cost. Besides uniqueness, another sufficient condition ensuring the optimality of a Pontryagin solution is the convexity of the functional $c'x(t_f)$ with respect to $u \in \mathcal{U}$.

Note that if $u^o(t)$ lies in the vertices of $\mathcal{U}$, then an *admissible switching signal* (i.e., $\sigma(t) \in \{1, 2, \ldots, N\}$ for almost all t) $\sigma^o(t)$ can be constructed as follows:

$$
\begin{aligned}
\dot{x}^o(t) &= A_{\sigma^o(t, x_0)} x^o(t), \\
-\dot{\pi}^o(t) &= A'_{\sigma^o(t, x_0)} \pi^o(t), \\
\sigma^o(t, x_0) &= \underset{i \in \mathcal{I}}{\operatorname{argmin}} \{\pi^{o'}(t) A_i x^o(t)\}
\end{aligned}
$$

with boundary conditions $x^o(0) = x_0$, $\pi^o(t_f) = c$, and $J(x_0, x^o, \sigma^o) = \pi^{o'}(0)x_0$.

Remark 8.3. Another interesting interpretation of the cost relies on the theory of Markov jump linear systems. Indeed, note that the state equation (7.2) can be written as follows:

$$\dot{x}_i(t) = \eta_{i,\sigma(t)} x_i(t) + \mu \sum_{j \neq i} \lambda_{ij} x_j(t), \tag{8.27}$$

where $\eta_{i,\sigma(t)} = \rho_{i,\sigma(t)} + 2\mu - \delta$ and $\lambda_{i,j} = m_{i,j}, i \neq j, \lambda_{i,i} = -2$. Note that matrix $\mu\Lambda$ with $\Lambda = \{\lambda_{i,j}\}$ is a stochastic matrix that can be considered as the infinitesimal transition matrix of the Markov jump linear system

$$\dot{\xi} = 0.5 \eta_{i,\sigma(t)} \xi. \tag{8.28}$$

Moreover, $\sum_{i=1}^{n} x_i(t) = E[\xi^2(t)]$. Minimizing $\sum_{i=1}^{n} x_i(t)$ is then equivalent to minimizing the variance of the stochastic process $\xi(t)$. In fact, if $\lim_{t \to \infty} E[\xi^2(t)] = 0$, then system (8.28) is stable in the mean-square sense.

It was explained in previous chapter that there is no general consensus among clinicians on the optimal time to change therapy in HIV infection to avoid virologic failure. The clinical trial in [228] suggested that proactive switching and alternation of antiretroviral regimens could extend the overall long-term effectiveness of the available therapies. Then, the optimal problem is formulated to mitigate viral escape as follows.

BOX 8.1

Problem 8.2. Given the 4-variant model given by system (7.2), the model can be rewritten as

$$\dot{x}(t) = \left(R_{\sigma(t)} - \delta_V I\right)x(t) + \mu M_u x(t), \qquad (8.29)$$

where $M_u := [m_{ij}]$ and $R_{\sigma(t)} := diag\{\rho_{i,\sigma(t)}\}$, $\rho_{i,\sigma(t)}$ is the replication rate for viral genotype i and therapy combination σ, μ represents the mutation rate, δ_V is the viral clearance, and $m_{i,j} \in \{0, 1\}$ represents the genetic connections between genotypes, namely, $m_{i,j} = 1$ if and only if it is possible for genotype j to mutate into genotype i [187].

The optimal control problem is to find for a fixed period of treatment t_f the optimal switching signal between two HAART regimens to minimize the total viral load represented by the performance criterion (8.23).

Motivated by the HIV treatment application stated in Problem 8.2, solutions based on the Pontryagin principle for particular subclasses of model (8.29) are developed. These subclasses are restricted to $N = 2$, which means there are only two available treatments. In this case, the optimal control (8.26) may be a sliding mode when the decision variable $\gamma(t)$ vanishes on a nontrivial interval, where

$$\gamma(t) := \pi'(t)(A_1 - A_2)x(t). \qquad (8.30)$$

Invariant Subspaces

One problem of interest is when x and π evolve on a particular invariant subspace as in the following lemma.

Lemma 8.1. *Consider the dynamic system (8.8) with initial condition $x(0) = x_0$, $N = 2$, and cost function $J = c'x(t_f)$. Suppose there exist $A_\alpha := \alpha A_1 + (1 - \alpha)A_2$ ($\alpha \in (0, 1)$) and tall matrices $v_x, w_x, v_\pi, w_\pi \in \mathbb{R}^{n \times m}$ (for any $m \leq n$) such that:*

(i) v'_x is a left invariant of A_α,

(ii) $v'_x x_0 = 0$,

(iii) v_π is a right invariant of A_α,

(iv) $c'v_\pi = 0$,

(v) $A_1 - A_2 = w_x v'_x + v_\pi w'_\pi$.

Then there exists a Pontryagin solution over $t \in [0, t_f]$ that is a sliding mode.

Proof. We state that

$$
\begin{aligned}
x(t) &= e^{A_\alpha t} x_0, \\
\pi(t) &= e^{A'_\alpha (t_f - t)} c',
\end{aligned}
$$

with the equivalent control $u_1(t) = \alpha$, $u_2(t) = (1 - \alpha)$, satisfies the requirements of Definition 8.12. Clearly, this is a valid solution to the state and costate equations, and it remains to show that it satisfies the restrictions on $\gamma(t)$. However, because of the invariant subspace assumptions, it is always true that the state lives in the invariant subspace $v'_x x(t) = 0$. Similarly, the costate always evolves within the invariant subspace $\pi(t)' v_\pi = 0$. From condition (v) and (8.30) it is shown that $\gamma(t) = 0$ for all $t \in [0, t_f]$. $\square$

Generalized Symmetry Case for Viral Mutation

An important case where the invariant subspace assumptions of Lemma 8.1 apply is a particular kind of symmetry. In fact, the result applies also to a certain generalization of this symmetry as shown in Fig. 8.1. To this end, we consider a generalized transposition assumption.

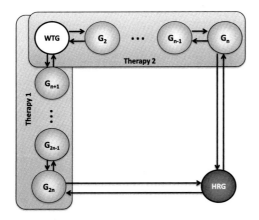

■ **FIGURE 8.1 General permutation case.** This mutation arrangement has the wild-type genotype that need to mutate several times to be resistant to therapy.

Definition 8.13. A matrix $T \in \mathbb{R}^{n \times n}$ is called a (generalized) transposition if there exists a matrix $v \in \mathbb{R}^{n \times m}$ such that

$$
T = I - vv' \quad \text{and} \tag{8.31}
$$

$$v'v = kI \text{ for some } k > 1. \tag{8.32}$$

Note, for example, that with $k = 2$ in Definition 8.13, T is precisely a transposition with $T^2 = I$.

Assumption 8.1. The matrix A_2 may be obtained as a (generalized) transposition of A_1, that is, there exists a transposition T such that

$$A_2 = T A_1 T. \tag{8.33}$$

The following lemma shows how this assumption may be used to construct an invariant subspace.

Lemma 8.2. *Under Assumption 8.1, v' is a basis for a left-invariant subspace of $A_\alpha = \alpha A_1 + (1 - \alpha) A_2$; in particular, $v' A_\alpha = \alpha (v' A_1 v) v'$ with $\alpha = \frac{k-1}{k}$.*

Proof. Using (8.33), A_α can be rewritten as

$$A_\alpha = \alpha A_1 + (1 - \alpha) T A_1 T.$$

Substituting (8.31) into A_α, we obtain

$$A_\alpha = \alpha A_1 + (1 - \alpha)(A_1 - A_1 vv' - vv' A_1 + vv' A_1 vv').$$

Multiplying v' from the left and using (8.32), we derive

$$v' A_\alpha = \alpha v' A_1 + (1 - \alpha)(k - 1)(v' A_1 vv' - v' A_1).$$

Then using $\alpha = \frac{k-1}{k}$, we obtain

$$v' A_\alpha = \alpha (v' A_1 v) v'. \qquad \square$$

We now extend this result to show how it may be used to find optimal switching controls.

Theorem 8.7. *Suppose that the following conditions are satisfied:*

1. *Assumption 8.1;*
2. *The initial conditions and cost satisfy $v' x_0 = 0$ and $v' c = 0$, with v as in Definition 8.13; and*
3. *A_1 and A_2 are symmetric.*

Then a Pontryagin solution is given by the trajectory along the plane $v' x(t) = 0$ with dynamical matrix A_α.

Proof. The proof follows directly from Lemmas 8.1 and 8.2 and the fact that the left- and right-invariant subspaces are equivalent for the symmetric matrix A_α. Condition (v) of Lemma 8.1 can be established as follows:

$$
\begin{aligned}
A_1 - A_2 &= A_1 - (I - vv')A_1(I - vv') \\
&= v\left(v'A_1 - \frac{1}{2}v'A_1vv'\right) + \left(A_1v - \frac{1}{2}vv'A_1v\right)v' \\
&= vw' + wv',
\end{aligned}
$$

where $w = A_1v - \frac{1}{2}vv'A_1v$. $\qquad\square$

Note that as well as the matrix version of symmetry, $A_1' = A_1$, there is also a permutation symmetry. In particular, T is now idempotent, and the dynamics under the two treatment options are identical apart from relabeling some state variables. Furthermore, the conditions that the initial conditions and costate start on the invariant subspace are equivalent to equal weighting and initial conditions on state variables that are transposed under T. Therefore, certain kinds of symmetric problems generically admit Pontryagin solutions that are a sliding mode. Next, a specific class of systems for which this solution is optimal is derived.

General Solution for a 4-Variant Model

Consider system (7.2), 4 states, and two treatment options rewritten as follows:

$$
A_\sigma = \begin{bmatrix} \lambda_1 & 0 & 0 & 0 \\ 0 & \lambda_{2\sigma} & 0 & 0 \\ 0 & 0 & \lambda_{3\sigma} & 0 \\ 0 & 0 & 0 & \lambda_4 \end{bmatrix} + \mu \begin{bmatrix} 0 & 1 & 1 & 0 \\ 1 & 0 & 0 & 1 \\ 1 & 0 & 0 & 1 \\ 0 & 1 & 1 & 0 \end{bmatrix}.
$$

Remark 8.4. In reality, HIV treatments are a combination of multiple (usually three or more) individual drugs, each of which has independent susceptibility to viral mutation. Even if considering only two classes of treatment, a full order model for this would have up to $2^{(3\times2)} = 64$ (or possibly more) viral strains. The 4-variant model is a significant simplification of complex mutation dynamics in real HIV behavior, which is more amenable to analysis. $\qquad\square$

Assumption 8.2. $\lambda_{21} > 0$, $\lambda_{22} < 0$, $\lambda_{31} < 0$, $\lambda_{32} > 0$.

In addition, we assume the following symmetry.

Assumption 8.3. $\lambda_{21} - \lambda_{22} + \lambda_{31} - \lambda_{32} = 0$.

Using Assumptions 8.2 and 8.3, $\Delta A = A_1 - A_2$ can be rewritten as follows:

$$\Delta A = (\lambda_{21} - \lambda_{22})\bar{J},$$

where $\bar{J} = diag\,(0, 1, -1, 0)$. Since $\lambda_{21} - \lambda_{22} > 0$, the normalized decision function is defined as $\bar{\gamma}(t) = \pi(t)'\bar{J}x(t)$, which takes the form

$$\bar{\gamma}(t) = \pi_2(t)[x_2(t) - x_3(t)] + x_3(t)[\pi_2(t) - \pi_3(t)]. \tag{8.34}$$

Moreover, from the structure of A_1 and A_2 it is possible to conclude that

$$\begin{aligned}\dot{\bar{\gamma}}(t) &= \mu[\pi_2(t) - \pi_3(t)][x_1(t) + x_4(t)] \\ &\quad -\mu[x_2(t) - x_3(t)][\pi_1(t) + \pi_4(t)].\end{aligned} \tag{8.35}$$

The following lemma, which can be proven directly from (8.34) and Assumption 8.2, is useful to characterize the optimal solution. We further introduce the sign function $\mathcal{S}_{gn}$ by $\mathcal{S}_{gn}[v] = 1$ if $v > 0$, $\mathcal{S}_{gn}[v] = -1$ if $v < 0$, and $\mathcal{S}_{gn}[0] = 0$.

Lemma 8.3. *Under Assumption 8.2, the following conditions hold for any Pontryagin solution $x(t)$, $\pi(t)$, $u(t)$:*

$$\begin{aligned}&\left\{\mathcal{S}_{gn}[x_2(t) - x_3(t)] = \mathcal{S}_{gn}[\pi_2(t) - \pi_3(t)]\right\} \\ &\quad\Longrightarrow \left\{\mathcal{S}_{gn}[\bar{\gamma}(t)] = \mathcal{S}_{gn}[x_2(t) - x_3(t)]\right\}, \\ &\left\{\mathcal{S}_{gn}[x_2(t) - x_3(t)] = -\mathcal{S}_{gn}[\pi_2(t) - \pi_3(t)]\right\} \\ &\quad\Longrightarrow \left\{\mathcal{S}_{gn}[\dot{\bar{\gamma}}(t)] = \mathcal{S}_{gn}[\pi_2(t) - \pi_3(t)]\right\}, \\ &\mathcal{S}_{gn}[\dot{x}_2(t) - \dot{x}_3(t)] = \mathcal{S}_{gn}[\alpha - u_1(t)], \\ &\mathcal{S}_{gn}[\dot{\pi}_2(t) - \dot{\pi}_3(t)] = \mathcal{S}_{gn}[u_1(t) - \alpha].\end{aligned}$$

To characterize the sliding modes, it is necessary to find a suitable convex combination of the matrices A_1, A_2 as follows.

Lemma 8.4. *Consider any time interval, $[t_1, t_2]$ and suppose that $x_2(t_1) = x_3(t_1)$, $\pi_2(t_2) = \pi_3(t_2)$. Under Assumption 8.3, there is a Pontryagin solution over $[t_1, t_2]$ such that $x_2(t) = x_3(t)$, $\pi_2(t) = \pi_3(t)$ with $\alpha = \frac{\lambda_{32} - \lambda_{22}}{\lambda_{32} - \lambda_{22} + \lambda_{21} - \lambda_{31}}$.*

Proof. Given the structure of $\bar{J}$, this can clearly be written as $\bar{J} = vw' + wv'$ where $w = \frac{1}{2}[0, 1, 1, 0]'$ and $v = [0, 1, -1, 0]'$. It suffices to show that the variables $x_2(t) - x_3(t)$ and $\pi_2(t) - \pi_3(t)$ obey autonomous differential equations. Indeed,

$$\dot{x}_2(t) - \dot{x}_3(t) = \alpha(\lambda_{21} - \lambda_{31})x_2(t) + (1 - \alpha)(\lambda_{22} - \lambda_{32})x_3(t),$$

where $\alpha(\lambda_{21} - \lambda_{31}) = (1 - \alpha)(\lambda_{22} - \lambda_{32})$, so that letting $r = \frac{\lambda_{21}\lambda_{32} - \lambda_{22}\lambda_{31}}{\lambda_{32} - \lambda_{22} + \lambda_{21} - \lambda_{31}}$, it follows that $\dot{x}_2(t) - \dot{x}_3(t) = r(x_2(t) - x_3(t))$ and $\dot{\pi}_2(t) - \dot{\pi}_3(t) = -r(\pi_2(t) - \pi_3(t))$. $\square$

For the future proof of optimality, the positive invariance of certain regions is established for any Pontryagin solutions to the optimal control problem.

Lemma 8.5. *With respect to any Pontryagin solution and subject to Assumptions 8.2–8.3, the following regions are positively invariant:*

$$\mathcal{R}_{+1} := \{(x, \pi) : \bar{\gamma} > 0, x_2 - x_3 \leq 0, \pi_2 - \pi_3 \geq 0\},$$
$$\mathcal{R}_{+2} := \{(x, \pi) : \bar{\gamma} < 0, x_2 - x_3 \geq 0, \pi_2 - \pi_3 \leq 0\}.$$

Proof. This result is proved for $\mathcal{R}_{+1}$ only as the other case follows similarly. Note that $\bar{\gamma} > 0$ implies that $u_1 = 0$, and therefore

$$\dot{x}_2 - \dot{x}_3 = \lambda_{22}x_2 - \lambda_{32}x_3 < 0,$$
$$\dot{\pi}_2 - \dot{\pi}_3 = -\lambda_{22}\pi_2 + \lambda_{32}\pi_3 > 0.$$

Furthermore, by (8.35), $\dot{\bar{\gamma}} \geq 0$ in this region. $\square$

Now define $k_1 = \text{argmin}\{x_2(0), x_3(0)\}$, $k_2 = \text{argmin}\{c_2, c_3\}$, and

$$T_1 = \underset{t \geq 0}{\text{argmin}} : [0 \ 1 \ -1 \ 0]e^{A_{k_1}t}x(0) = 0,$$
$$T_2 = \underset{t \leq t_f}{\text{argmax}} : [0 \ 1 \ -1 \ 0]e^{-A_{k_2}(t-t_f)}c = 0.$$

Note that, thanks to the definition of k_1, k_2 and the monotonicity conditions of $x_2(t) - x_3(t)$ and $\pi_2(t) - \pi_3(t)$, the time instants T_1 and T_2 are well defined and unique. Clearly, by definition $x_2(T_1) = x_3(T_1)$ and $\pi(T_2) = \pi_3(T_2)$. Now the main result can be derived for the four-variant model (recall that $\sigma = 1$ means $u_1 = 1$, $\sigma = 2$ means $u_1 = 0$, and a sliding mode correspond to the singular arc associated with $u_1 = \alpha$).

Theorem 8.8. *Long Horizon Case*

Let Assumptions 8.2 and 8.3 be satisfied, let $\mu > 0$, and assume that $T_1 \leq T_2$. Then the optimal control associated with the initial state $x(0)$ and cost $c'x(t_f)$ is given by $\sigma(t) = k_1$, $t \in [0, T_1]$, and $\sigma(t) = k_2$, $t \in [T_2, t_f]$. For $t \in [T_1, T_2]$, the optimal control is given by the trajectory along the plane $x_2 = x_3$ with dynamical matrix $A_\alpha = \alpha A_1 + (1 - \alpha)A_2$ and α as given in Lemma 8.2.

Proof. The candidate optimal solution triple is defined as follows:

$$
\begin{aligned}
u_1(t) &= k_1, & t &\in [0, T_1], \\
u_1(t) &= \alpha, & t &\in [T_1, T_2], \\
u_1(t) &= k_2, & t &\in [T_2, t_f]; \\
x(t) &= e^{A_{k_1} t} x(0), & t &\in [0, T_1], \\
x(t) &= e^{A_\alpha (t-T_1)} x(T_1), & t &\in [T_1, T_2], \\
x(t) &= e^{A_{k_2} (t-T_2)} x(T_2), & t &\in [T_2, t_f]; \\
\pi(t) &= e^{A_{k_2} (t_f - t)} c, & t &\in [T_2, t_f], \\
\pi(t) &= e^{A_\alpha (T_2 - t)} \pi(T_2), & t &\in [T_1, T_2], \\
\pi(t) &= e^{A_{k_1} (T_1 - t)} \pi(T_1), & t &\in [0, T_1].
\end{aligned}
$$

First, consider the time interval $t \in [T_1, T_2]$. By definition, $x_2(T_1) = x_3(T_1)$ and $\pi_2(T_2) = \pi_3(T_2)$. By Lemma 8.4 we can obtain $x_2(t) = x_3(t)$ and $\pi_2(t) = \pi_3(t)$ in the interval $[T_1, T_2]$, and therefore the Pontryagin conditions are satisfied in this interval.

For the remaining two intervals $t \in [0, T_1]$ $(i = 1)$ and $t \in [T_2, t_f]$ $(i = 2)$, consider $\bar{\gamma}$ (8.34) and $\dot{\bar{\gamma}}$ (8.35). Now, $\bar{\gamma}(T_1) = \bar{\gamma}(T_2) = 0$, and, for $t \in [0, T_1]$ or $t \in [T_2, t_f]$, we have:

$$
\dot{x}_2(t) - \dot{x}_3(t) = \lambda_{2k_i} x_2(t) - \lambda_{3k_i} x_3(t) =
\begin{cases}
> 0, & k_i = 1, \\
< 0, & k_i = 2,
\end{cases}
$$

$$
\dot{\pi}_2(t) - \dot{\pi}_3(t) = -\lambda_{2k_i} \pi_2(t) + \lambda_{3k_i} \pi_3(t) =
\begin{cases}
< 0, & k_i = 1, \\
> 0, & k_i = 2.
\end{cases}
$$

This means that, for $t \in [0, T_1]$ or $t \in [T_2, t_f]$, we have:

$$
x_2(t) - x_3(t) =
\begin{cases}
< 0, & k_i = 1, \\
> 0, & k_i = 2,
\end{cases}
$$

$$
\pi_2(t) - \pi_3(t) =
\begin{cases}
> 0, & k_i = 1, \\
< 0, & k_i = 2.
\end{cases}
$$

Therefore, within either time interval, the sign of $\dot{\bar{\gamma}}$ is uniform:

$$
\dot{\bar{\gamma}}(t) =
\begin{cases}
> 0, & k_i = 1, \\
< 0, & k_i = 2.
\end{cases}
$$

Since $\bar{\gamma}(T_i) = 0$ for $i = 1, 2$, it follows that the sign of $\bar{\gamma}$ is uniform within either time interval and takes the required sign to satisfy the Pontryagin conditions. It has to be shown that the proposed Pontryagin solution $u(t), x(t), \pi(t)$ is unique so that optimality follows. To do this, we consider other candidate Pontryagin solutions and produce a contradiction, thereby establishing that the proposed solution is optimal. Making reference to Figs. 8.2A–B that correspond to the case $x_2(0) < x_3(0)$, the case $x_2(0) = x_3(0)$ can be understood in a simple way via the same figures with $T_1 = 0$, whereas the case $x_2(0) > x_3(0)$ can be obtained by symmetry.

Case (a): $\pi_2(0) - \pi_3(0) \leq 0$. Since $x_2(0) < x_3(0)$, it follows that $\bar{\gamma}(0) < 0$ and $u_1(0) = 1$. Then, initially, $x_2(t) - x_3(t)$ increases, and $\pi_2(t) - \pi_3(t)$ decreases. Therefore there is an initial time interval $[0, \tau_1]$ wherein $\bar{\gamma}(t) < 0$, $\dot{\pi}_2 < \dot{\pi}_3$, and therefore $\pi_2(t) < \pi_3(t)$. These conditions persist unless $x_2 = x_3$ is reached. Nevertheless, at this point, trajectories may enter the positively invariant region $\mathcal{R}_{+2}$. Therefore, in this case, $\bar{\gamma} < 0$ and $\pi_2 < \pi_3$ must be maintained over the entire time interval. However, this contradicts either (i) $\pi_2(t_f) - \pi_3(t_f) = c_2 - c_3$ if $c_2 > c_3$ or (ii) the long horizon assumption on the existence of T_2 if $c_2 < c_3$, which yields $\pi_2(T_2) = \pi_3(T_2)$.

Case (b): $\pi_2(0) - \pi_3(0) \geq 0$, $\bar{\gamma}(0) > 0$. In this case, trajectories start in $\mathcal{R}_{+1}$ and therefore remain in this region. In a similar manner to case (a), this produces a contradiction with respect to either the final value of the costate or the existence of T_2.

Case (c): $\pi_2(0) - \pi_3(0) > 0$, $\bar{\gamma}(0) \leq 0$. Note that if $\bar{\gamma}(0) = 0$, then at some small time, later trajectories are effective in case (b), and the contradiction there holds. Therefore $\bar{\gamma}(t)$ remains negative until some $t = \tau_1 > 0$. Suppose $\tau_1 < T_1$; then by the definition of T_1, $x_2(\tau_1) < x_3(\tau_1)$, and therefore we must have $\bar{\gamma}(\tau_1) = 0$ and $\pi_2(\tau_1) > \pi_3(\tau_1)$. Then just after $t = \tau_1$, trajectories enter $\mathcal{R}_{+1}$ and remain there, and as in cases (a), (b), this leads to a contradiction. Suppose instead that $\tau_1 > T_1$. Then for $t \in (T_1, \min\{\tau_1, T_2\})$, we must have $x_2(t) > x_3(t)$ and $\bar{\gamma}(t) < 0$. This implies that $\pi_2(t) < \pi_3(t)$, and therefore trajectories are in $\mathcal{R}_{+2}$, and we get a similar contradiction as before. Therefore the only case that does not lead to a contradiction is that $\bar{\gamma}(t)$ reaches zero precisely at $t = T_1$, that is, $x_2(T_1) = x_3(T_1)$ and $\pi_2(T_1) = \pi_3(T_1)$ as in the proposed solution.

By reversing the same logic, we can show that the final time interval must be of the form postulated, that is, $x_2(T_2) = x_3(T_2)$ and $\pi_2(T_2) = \pi_3(T_2)$. Finally, the only Pontryagin solution in the interval $[T_1, T_2]$ with $x_2(T_i) = x_3(T_i)$ and $\pi_2(T_i) = \pi_3(T_i)$ is $u_1(t) = \alpha$. $\qquad \Box$

Even though in practice the horizon length t_f may often be large enough to guarantee that $T_1 \leq T_2$, for completeness, we also consider the *small horizon* case.

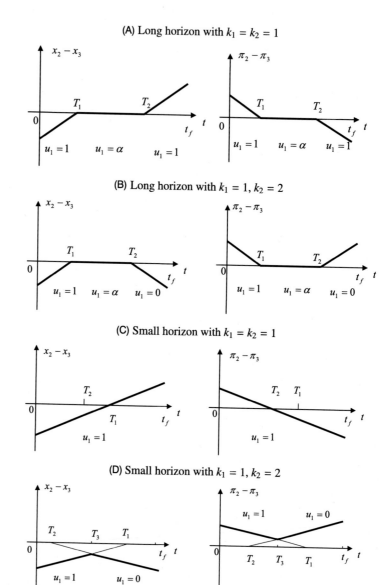

■ **FIGURE 8.2 Long and small horizon cases.** The differences $x_2 - x_3$ and $\pi_2 - \pi_3$ are plotted correspondingly to long and short horizons.

Theorem 8.9. *Small Horizon Case*

Let Assumption (8.2) be satisfied and assume that $\mu \neq 0$ and $0 < T_2 \leq T_1 < t_f$. Then the optimal control associated with the initial state $x(0)$ and cost $c'x(t_f)$ is given as follows:

(i) *If $k_1 = k_2$, then $\sigma(t) = k_1$, $t \in [0, t_f]$;*

(ii) *otherwise, if $k_1 \neq k_2$, then $\sigma(t) = \left\{ \begin{array}{ll} k_1: & t \in [0, T_3] \\ k_2: & t \in [T_3, t_f] \end{array} \right\}$, where $T_3 \in$*

$[T_2, T_1]$ *is such that, for $t = T_3$,*

$$g(t) := x(0)' e^{A_{k_1}(t)} \bar{J} e^{-A_{k_2}(t - t_f)} c = 0.$$

Proof. The proof follows similar steps of Theorem 8.8. Therefore, we omit many details. The proposed control law should satisfy the Pontryagin conditions (8.9). Consider case (i) $k_1 = k_2$ (see Fig. 8.2C). Also, take $k_1 = 1$, with the other case following by symmetry. Our candidate solution is

$$\begin{aligned} \pi(t) &= e^{A_1(t_f - t)} c, & t \in [0, t_f], \\ x(t) &= e^{A_1 t} x(0), & t \in [0, t_f], \end{aligned}$$

and $u_1(t) = 1$ for all t. It remains to confirm that this satisfies the decision condition (8.26). Note that by the definition of T_2, $x_2(t) - x_3(t)$ changes sign only at $t = T_2$. Similarly, $\pi_2(t) - \pi_3(t)$ changes sign only at $t = T_1$. For $t \in [\max\{0, T_2\}, \min\{T_1, t_f\}]$, both $x_2(t) - x_3(t)$ and $\pi_2(t) - \pi_3(t)$ are both negative, and therefore $\bar{\gamma}(t) < 0$ as required. Further, two regions may need to be considered, $[0, T_2]$ and $[T_1, t_f]$. In these regions, $\dot{\bar{\gamma}}(t) > 0$ and $\dot{\bar{\gamma}}(t) < 0$ must hold, respectively, and therefore $\bar{\gamma}(t) < 0$ for the entire interval $t \in [0, t_f]$.

Now consider case (ii) $k_1 \neq k_2$ (see Fig. 8.2C). Having $k_1 = 1$ and $k_2 = 2$ with the other cases following similarly, we first show that T_3 exists. Note that $g(T_2)$ is negative and $g(T_1)$ is positive. The uniqueness follows since

$$g'(t) := x(0)' e^{A_1(t)} (A_1 \bar{J} - \bar{J} A_2) e^{-A_2(t - t_f)} c \geq 0,$$

where the inequality follows since, for our case, $(A_1 \bar{J} - \bar{J} A_2)$ is diagonal and nonnegative. Our candidate optimal solution is therefore

$$\begin{aligned} \pi(t) &= e^{A_2(t_f - t)} c, & t \in [T_3, t_f], \\ \pi(t) &= e^{A_1(T_3 - t)} e^{A_2(t_f - T_3)} c, & t \in [0, T_3]; \\ x(t) &= e^{A_1 t} x(0), & t \in [0, T_3], \\ x(t) &= e^{A_2(t - T_3)} e^{A_1 T_3} x(0), & t \in [T_3, t_f]; \\ u_1(t) &= 1, & t \in [0, T_3), \\ u_1(t) &= 2, & t \in (T_3, t_f]. \end{aligned}$$

By the definition of T_3 the candidate $\bar{\gamma}(t)$ swaps sign at $t = T_3$. Now we need to show that the associated Pontryagin triple is unique. Note that by the definitions of T_1 and T_2, for any Pontryagin solution, $x_2(t) < x_3(t)$ for $t \in [0, T_1)$ and $\pi_2(t) > \pi_3(t)$ for $t \in (T_2, t_f]$.

Now consider three cases. Case (a): $\pi_2(0) \leq \pi_3(0)$. This leads to $\bar{\gamma} < 0$, and therefore $\pi_2 - \pi_3$ decreases. It can be shown that this situation persists, and therefore there is a contradiction with the terminal condition $\pi(t_f) = c$. Case (b): $\pi_2(0) > \pi_3(0)$ and $\bar{\gamma}(0) > 0$. This is part of the set $\mathcal{R}_{+1}$, which is positively invariant, and yields a valid solution only when $t_f < T_3 < T_1$. Case (c): $\pi_2(0) > \pi_3(0)$ and $\bar{\gamma}(0) < 0$. This yields $\dot{\bar{\gamma}}(0) > 0$ and the solution postulated wherein $T_3 < T_1 < t_f$.

Reversing the argument to work backward from $t = t_f$ gives the solution postulated, including possible cases where T_3 exists but is outside the range $(0, t_f)$. □

Remark 8.5. Theorem 8.9 implicitly includes the cases where one or more of T_1, T_2, and T_3 are outside the range $(0, t_f)$.

Numerical Simulations. HIV treatments are designed to require the accumulation of three or more resistance mutations before the appearance of a fully resistant variant. This would give a complex scenario with a much higher degree of complexity. To illustrate the optimal control policies, we employ the example from [247] using a 4-variant, 2-drug treatment model with the initial condition vector $x = [10^3, 5, 0, 10^{-5}]$ and the symmetric cost function weighting as $c = [1, 1, 1, 1]'$. The viral clearance rate is $\delta_V = 0.24$ day^{-1}, which corresponds to a half-life of less than 3 days [202]. Viral mutation rates are of the order of $\mu = 10^{-4}$. The various replication rates are described in Table 8.1, where numbers are of course idealized; however, the general principles are based on [187,247].

Table 8.1 Symmetric replication rates for viral variants

Variant	Therapy 1	Therapy 2
Wild type (x_1)	$\rho_{1,1} = 0.05$	$\rho_{1,2} = 0.05$
Genotype 1 (x_2)	$\rho_{2,1} = 0.27$	$\rho_{2,2} = 0.05$
Genotype 2 (x_3)	$\rho_{3,1} = 0.05$	$\rho_{3,2} = 0.27$
HR Genotype (x_4)	$\rho_{4,1} = 0.27$	$\rho_{4,2} = 0.27$

Using Theorem 8.8, therapy 2 is used for $t < T_1 = 12.2$ days, then therapies alternate with high frequency as illustrated in Fig. 8.3A. The wild-type virus is attenuated to undetectable levels (less than 50 copies/ml). The highly resistant genotype grows slowly, which induces the final viral escape. Fig. 8.3A reveals how the costate dynamics are similar to the state but in reverse time. Using a switch on virological failure strategy, the therapy is changed after 9 months (when viral load $\geq$ 1000 copies/ml). Therefore the population of the resistant genotype is large enough and cannot be contained by the second therapy; see Fig. 8.3B. In contrast, proactive switching may reduce viral load to very low levels during the whole treatment, 100 copies/ml, promoting a larger delay in the viral escape. Note that an open-loop alternating strategy and the optimal control present close performance for this example. This means that a periodic oscillating strategy might be effective in postponing viral escape without requiring a detailed model, high computational time, and full state measurements.

Remark 8.6. Optimal trajectories are associated with chattering switching laws, which are of course not realistically applicable for HIV treatment. However, this theoretical result provides an important insight since it clarifies when the therapies have to be switched more frequently in order to better control viral load.

8.5 RESTATEMENT AS AN OPTIMIZATION PROBLEM

In some situations the optimal control problem may be restated as a nonlinear optimization problem, which can be considered as an easier choice than the solution of the HJE. For this purpose, instead of finding when the system will switch, it may be considered that the system will switch, but now the optimal length of the switches is requested. Using the system Σ_A, it is taken a constant number of switches for the period t_f. The state is known at every moment using the exponential matrix, and we can obtain

$$x(\tau_1) = e^{A_1 \Delta \tau_1} x(0). \tag{8.36}$$

The problem can be expressed in time t_f as

$$x(t_f) = \prod_{i=1}^{N_f} e^{A_j \Delta \tau_i} x(0), \tag{8.37}$$

where N_f is the number of switches on the interval $[t_0, t_f]$, $j = 1$ when i is odd, and otherwise $j = 2$. In addition, the following restriction must be satisfied:

$$t_f = \sum_{i}^{N_f} \Delta \tau_i. \tag{8.38}$$

Now, if the cost function (8.23) is taken, then this problem can be rewritten in a form where the cost function depends just of $\Delta \tau_i$:

$$J(t_f) = c' \prod_{i=1}^{N_f} e^{A_j \Delta \tau_i} x(0). \tag{8.39}$$

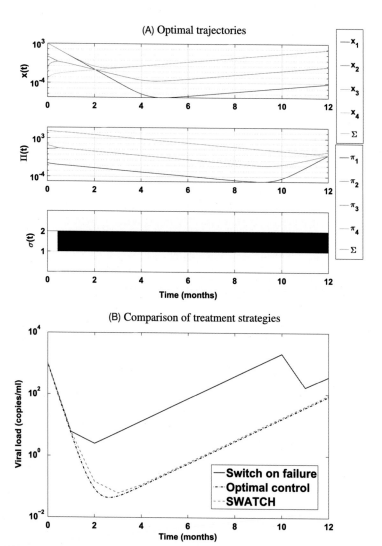

■ FIGURE 8.3 **Treatment strategies to Mitigate Viral escape.** Top panel presents optimal trajectories and the corresponding switching rules. Bottom panel compares different treatment strategies.

Lemma 8.6. *Let $\mu = 0$ for system (7.2). Then the optimal switching signal is described by a single switch with duration*

$$T_2 = \frac{1}{2(\lambda_{21} - \lambda_{22})} \ln \frac{x_3(t_f)^{\sigma=1}}{x_2(t_f)^{\sigma=1}}. \tag{8.40}$$

Proof. Assuming that the mutation rate μ is 0, the matrices A_i are diagonal. Since diagonal matrices commute ($A_1 A_2 = A_2 A_1$), Eq. (8.39) reduces to

$$J(t_f) = c'e^{A_1 \sum_{i=1}^{N_f} \Delta\tau_i + A_2 \sum_{j=2}^{N_f} \Delta\tau_j} x(0), \qquad (8.41)$$

where $i \in$ even numbers and $j \in$ odd numbers. If the topology of the system $\Delta A = A_1 - A_2$ is used, then (8.41) can be rearranged as

$$
\begin{aligned}
J(t_f) &= c'e^{A_1(\sum_{i=1}^{N_f} \Delta\tau_i + \sum_{j=2}^{N_f} \Delta\tau_j) + \Delta A \sum_{j=1}^{N_f} \Delta\tau_j} x(0) \\
&= c'e^{\Delta A \sum_{j=2}^{N_f} \Delta\tau_j} x(t_f)^{\sigma=1},
\end{aligned}
$$

where $x(t_f)^{(\sigma=1)} = [x_1(t_f)^{\sigma=1}, x_2(t_f)^{\sigma=1}, x_3(t_f)^{\sigma=1}, x_4(t_f)^{\sigma=1}]'$ is the state vector at time t_f under the only effect of $\sigma = 1$. As a result of the commutation property, the order of the switches is not important for this case, and what is really important is how long $\sigma = 2$ is used independently of the order. For this purpose, we consider that $T_2 = \sum_{j=2}^{N_f} \Delta\tau_j$ as a variable with the only restriction that $T_2 \geq 0$. Computing the first derivative with respect to T_2, we have

$$\frac{dJ(t_f)}{dT_2} = c' \Delta A e^{\Delta A T_2} x_1(t_f). \qquad (8.42)$$

To find the minimum, we solve it for T_2:

$$T_2 = \frac{1}{2(\lambda_{21} - \lambda_{22})} \ln \frac{x_3(t_f)^{\sigma=1}}{x_2(t_f)^{\sigma=1}}.$$

To minimize the cost function $J(t_f)$ in system (7.2), it is required to use $\sigma = 2$ for the period T_2. This solution is nonunique, and for symmetric initial conditions, under Assumptions 8.2 and 8.3, the sliding-mode control $u_1(t) = \alpha$ is also optimal. $\quad\square$

Remark 8.7. For the case $\mu \neq 0$, the problem needs to be solved numerically using nonlinear optimization. However, because the problem is nonconvex, numerical solution is complex.

8.6 DYNAMIC PROGRAMMING FOR POSITIVE SWITCHED SYSTEMS

We describe a continuous-time model for treatment of the viral mutation given in (7.2). In practice, measurements can only reasonably be made infrequently. For simplicity, we consider a regular treatment interval τ, during which the treatment is fixed. If $k \in \mathbb{N}$ denotes the number of intervals since $t = 0$, then

$$x(k+1) = A_{\sigma(k)} x(k) \qquad (8.43)$$

is defined for all $k \in \mathbb{N}$, where $x \in R^n$ is the state, $\sigma(k)$ is the switching sequence, and $x(0) = x_0$ is the initial condition. For (8.43) to be a positive system for any switching sequence, A_i, $i = 1, ..., N$, must be nonnegative matrices, that is, its entries are $a^i_{lj} \geq 0$ for all (l, j), $l \neq j$, $i = 1, 2, ..., N$. For each $k \in \mathbb{N}$,

$$\sigma(k) \in \{1, 2, ..., N\}. \tag{8.44}$$

Consider the following discrete-time cost function to be minimized over all admissible switching:

$$J = c'x(t_f) + \sum_{k=0}^{t_f - 1} q'_{\sigma(k)}x(k), \tag{8.45}$$

where $x(k)$ is a solution of (8.43) with switching signal $\sigma(k)$. The vectors c and q_i, $i = 1, 2, \cdots, N$, are assumed to be positive.

Theorem 8.10. *Discrete-time Optimal Control for Positive Switched Systems*

Let $\sigma^o(k, x_0) : [0, t_f] \times R^n_+ \to \mathcal{I} = \{1, ..., N\}$ be an admissible switching signal relative to x_0, and let $x^o(k)$ be the corresponding trajectory. Let $\pi^o(k)$ denote a positive vector solution of the system of difference equations

$$x^o(k + 1) = A_{\sigma^o(k)}x^o(k), \; x(0) = x_0,$$
$$\pi^o(k) = A'_{\sigma^o(k)}\pi^o(k + 1) + q_{\sigma^o(k)}, \; \pi(t_f) = c, \tag{8.46}$$
$$\sigma^o(k) = \operatorname*{argmin}_s\{\pi^o(k + 1)'A_sx^o(k) + q_sx^o(k)\}$$

with the boundary conditions $x^o(0) = x_0$ and $\pi^o(t_f) = c$. Then $\sigma^o(k, x_0)$ is an optimal switching signal relative to x_0.

Proof. We denote the optimal switching signal, the corresponding trajectory, and the optimal cost functional by $\sigma^o(k)$, $x^o(k)$, and $J(x_0, x^0, \sigma^0)$, respectively. Letting $u = \sigma(k)$ and $q(k, x, u) = q_{\sigma(k)}$, and using the Hamilton–Jacobi–Bellman equation for the discrete case, we can derive

$$\bar{V}(x, k) = \min_{u \in U}\{q(k, x, u) + \bar{V}(x(k + 1), k + 1)\}, \tag{8.47}$$

where, denoting the costate vector by $\pi(k)$, the general solution for this system is

$$\bar{V}(x(k), k) = \pi(k)'x(k). \tag{8.48}$$

Using Eqs. (8.43), (8.45), (8.47), and (8.48), we obtain the following system:

$$x^o(k+1) = A_{\sigma^o(k)}x^o(k) \, , \, x(0) = x_0,$$
$$\pi^o(k) = A'_{\sigma^o(k)}\pi^o(k+1) + q_{\sigma^o(k)} \, , \, \pi(t_f) = c, \quad (8.49)$$
$$\sigma^o(k) = \operatorname*{argmin}_{s}\{\pi^o(k+1)'A_s x^o(k) + q_s x^o(k)\}. \quad \square$$

Note that the discrete-time version needs also to be iterated forward, whereas the costate equation is iterated backward, both according to the coupling condition given by the switching rule σ. As a result, the problem is a two-point boundary value problem and cannot be solved using regular techniques. For this case, dynamic programming techniques can be used to solve numerically [237]. The discrete-time version results in a recursive equation, easily programmed for optimization. Before the presentation of dynamic programming algorithms, we establish an important property of the optimal value function $\bar{V}$.

Lemma 8.7. *For any k, the function $\bar{V}(x,k)$ is concave and positively homogeneous as a function of x.*

Proof. The fact that the function $\bar{V}(x,k)$ is positively homogeneous is obvious from (8.48). To prove the concavity, consider two initial states x_A and x_B and take any convex combination $x = \alpha x_A + \beta x_B$, $\alpha, \beta \geq 0$ and $\alpha + \beta = 1$. Let $\sigma^o(k)$ be the optimal sequence associated with initial condition x achieving the optimal cost J^o. Let $x_A(k)$ and $x_B(k)$ be the state sequences corresponding to $\sigma^o(k)$ and the initial states $x_A(0) = x_A$ and $x_B(0) = x_B$. By linearity of the system we have

$$x^o(k) = \alpha x_A(k) + \beta x_B(k).$$

Denote by J_A and J_B the (nonoptimal) costs associated with these sequences and denote by J_A^o and J_B^o the optimal costs with initial conditions x_A and x_B. In view of the linearity of the cost, we have

$$J^o = \alpha J_A + J_B \geq \alpha J_A^o + J_B^o.$$

This proves the concavity of $\bar{V}(x,0)$. The concavity for a generic k can be proved by dynamic programming arguments. $\square$

The previous lemma has several implications including the fact that given any convex combination (in a general polytope) of initial conditions, the best cost is achieved on a vertex. Without loss of generality, consider the case where $q = 0$, that is, there is a terminal cost only. Note that if this is not the case, a new variable $y(k)$ is introduced having the equation $y(k+1) = y(k) + q'x(k)$ and initial condition $y(0) = 0$, so that $J = c'x(t_f) + y(t_f)$.

In this way the original optimal control problem is reduced to the problem in which only the final cost $J = c'x(t_f)$ is considered. Given the initial condition $x(0)$, the optimal control problem turns out to be

$$\min_{i_{t_f}, i_{t_f-1}, \dots, i_1} c' A_{i_{t_f}} A_{i_{t_f-1}} \dots A_{i_1} x(0).$$

Let us recursively define the sequence of matrices

$$\Omega_0 = c,$$
$$\Omega_1 = [A'_1 \Omega_0 \ A'_2 \Omega_0 \ \dots \ A'_N \Omega_0] = [A'_1 c \ A'_2 c \ \dots \ A'_N c],$$
$$\vdots$$
$$\Omega_{k+1} = [A'_1 \Omega_k \ A'_2 \Omega_k \ \dots \ A'_N \Omega_k].$$

Then $\bar{V}(x, 0) = \min_i \Omega'_{t_f, i} x$, where $\Omega_{t_f, i}$ is the ith column of Ω_{t_f} and, in general,

$$\bar{V}(x, k) = \min_i \Omega'_{t_f-k, i} x(k). \tag{8.50}$$

At each step of the evolution the feedback strategy can be computed as

$$u(x(k)) = \operatorname*{argmin}_i \Omega'_{t_f-k, i} x(k),$$

that is, selecting the smallest component of the vector $\Omega'_{t_f-k} x(k)$.

The implementation of the strategy requires storing the columns of $\Omega'_{T-k} x(k)$. These number columns would be $1 + N + N^2 + N^3 + \dots + N^{t_f}$. This exponential growth can be too computationally demanding. Bellman called this difficulty as **"curse of dimensionality"**, since for high-dimensional systems, the number of high-speed storage locations become prohibitive. In general, many columns of the matrices Ω_k may be redundant and can be removed. This can be done by applying established dynamic programming methods as follows; see [248] for more detail.

Algorithm 1: Reverse Time Solution

Given $\Omega_{k,i}$, solve the linear programming (LP) problem

$$\mu_{k,i} = \min_{x:\, \Omega'_{k,\bar{i}}\, x \,\geq\, \bar{1}} \Omega'_{k,i} x, \tag{8.51}$$

where $\bar{1} = [1 \ 1 \ \dots \ 1]'$, and $\Omega_{k,\bar{i}}$ the matrix obtained from Ω_k by deleting the ith column. Then if $\mu_{k,i} \geq 1$, then the column $\Omega_{k,i}$ is redundant (and it should be eliminated from Ω_k). This means that, for each Ω_k, a "cleaned" version $\widehat{\Omega}_k$ of Ω_k can be generated, in which all the redundant columns are removed. Note that this elimination can be done while constructing the matrices Ω_k.

BOX 8.2 ALGORITHM 1: REVERSE TIME SOLUTION

1. For a finite step number t_f, suppose the initial condition is known for the state (x_0) and the final costate condition $\pi(t_f) = c$
2. Define $\Omega_{t_f}^{(1)} = c$ and set $k = t_f$
3. Compute the matrix

$$\widehat{\Omega}_k = [A_1' \Omega_{k-1}^{(1)} \ A_2' \Omega_{k-1}^{(1)} \ \cdots \ A_N' \Omega_{k-1}^{(1)}]$$

4. For each column i of $\widehat{\Omega}_k$
   i) Solve the LP (8.51) with Ω_k set to $\widehat{\Omega}_k$
   ii) If $\mu_{k,i} \geq 1$, then delete column i from $\widehat{\Omega}_k$
5. After examining all the columns, a reduced $\widehat{\Omega}_k$ is obtained, set $\Omega_k^{(1)} = \widehat{\Omega}_k$, set $k = k - 1$.
6. If $k \geq 0$, then return to (iii), otherwise continue
7. The optimal sequence will be given by

$$\sigma(k) = \operatorname*{argmin}_i \Omega_{k,i}'^{(1)} x_0$$

∎

Indeed, any redundant column of Ω_k necessarily produces only redundant columns in Ω_{k-1}. Then the procedure for the generation of a reduced representation $\Omega_k^{(1)}$ is achieved by performing the procedure described in Box 8.2. Therefore, although the exact solution in general is of exponential complexity, it may be computationally tractable for problems of reasonable dimension in terms of horizon and number of matrices. One way to further reduce the computational burden is to accompany the above algorithm (backward iteration) with its dual version (forward iteration).

Remark 8.8. A dual version of the algorithm may be constructed by taking the forward iterations

$$\Theta_0^{(1)} = x(0),$$
$$\widehat{\Theta}_{k+1}^{(1)} = [A_1 \Theta_k^{(1)} \ A_2 \Theta_k^{(1)} \ \cdots \ A_N \Theta_k^{(1)}].$$

Then the optimal feedback strategy can be computed as

$$\sigma(k) = \operatorname*{argmin}_i \Theta_{N-k,i}' c,$$

so that one can solve the LP problem

$$\nu_{k,i} = \min_{\pi:\, \Theta'_{k,\bar{i}}\, \pi\, \geq \bar{1}} \Theta'_{k,i}\pi,$$

where $\Theta_{k,\bar{i}}$ is the matrix obtained from Θ_k by deleting the ith column. In this case, if $\nu_{k,i} \geq 1$, then column i of Θ_k is redundant and may be removed.

Remark 8.9. For a given initial state x_0 and final cost vector c, we can combine both the reverse and forward time solutions to a midpoint (e.g., $t_f/2$) and finding $\min_{i,j} \Omega'^{(1)}_{t_f/2,i} \Theta^{(1)}_{t_f/2,j}$.

Algorithm 2: Box Constraint Algorithm

Algorithm 1 removes columns that are redundant for any x in R^n_+. This can be improved if tighter bounds are derived on $x(k)$, which apply independently of the switching sequence. If $A_{LB} \preceq A_i \preceq A_{UB}$ for all i, where bounds can be chosen as $A_{LB} = \min A_i$ and $A_{UB} = \max A_i$, then

$$A^k_{LB}x_0 \leq x(k) \leq A^k_{UB}x_0. \tag{8.52}$$

Therefore (8.51) can be replaced with the test

$$\mu_{k,i} = \min_{x,\alpha:\, \alpha \geq 0,\, \Omega_{k,\bar{i}}\, x \geq \alpha \bar{1},\, \beta_k} \Omega'_{k,i}x - \alpha, \tag{8.53}$$

where β_k represents inequality (8.52). If $\mu_{k,i} \geq 0$, then $\Omega_{k,i}$ is redundant.

Algorithm 3: Joint Forward/Backward Box Constraint Algorithm

Using a box constraint, the search space for Algorithm 1 is reduced. Remark 8.9 is applied to further improve the last algorithm. Instead of solving $t_f/2$ steps forward, then $t_f/2$ steps backward, and then combining sequences to get the optimal, backward–forward is solved step by step to make a tighter box constraint as in Box 8.3.

BOX 8.3 ALGORITHM 3: JOINT FORWARD/BACKWARD BOX CONSTRAINT

1. Initialize $\Omega^{(3)}_{t_f} = c$ and $\Theta^{(3)}_0 = x_0$, $s = 1$
   one step backward
2. Find

$$\widehat{\Omega}^{(3)}_{t_f-s} = [A'_1\Omega^{(3)}_{t_f-s+1} \quad A'_2\Omega^{(3)}_{t_f-s+1} \quad \cdots \quad A'_N\Omega^{(3)}_{t_f-s+1}]$$

3. For every ℓ, solve the LP given in (8.53) using the next tighter box constrained:

$$A_{LB}^{t_f-2s+1} x_{LB,s-1} \leq x_{t_f-2s+1} \leq A_{UB}^{t_f-2s+1} x_{UB,s-1}$$

where $x_{LB,s-1} = \min_\ell \Theta_{s-1,\ell}$ and $x_{UB,s-1} = \max_\ell \Theta_{s-1,\ell}$

4. Delete column $\widehat{\Omega}_{t_f-s,\ell}$ if $\mu_{t_f-s,\ell} \geq \alpha$

5. After examining all the columns, set $\Omega_{t_f-s}^{(3)} = \widehat{\Omega}_{t_f-s}^{(3)}$
   one step forward

6. Find

$$\widehat{\Theta}_s^{(3)} = [A_1 \Theta_{s-1}^{(3)} \quad A_2 \Theta_{s-1}^{(3)} \quad \cdots \quad A_N \Theta_{s-1}^{(3)}]$$

7. For every ℓ, remove the column $\Theta'_{s,\ell}$ and solve the LP given in (8.53) using the tighter box constrained:

$$A'^{t_f-2s}_{LB} \pi_{LB,t_f-s} \leq \pi_s \leq A'^{t_f-2s}_{UB} \pi_{UB,t_f-s}$$

where $\pi_{LB,t_f-s} = \min_\ell \Omega_{t_f-s,\ell}$ and $\pi_{UB,t_f-s} = \max_\ell \Omega_{t_f-s,\ell}$.

8. Delete column $\Theta_{s,\ell}$ if $\mu_{s,\ell} \geq \alpha$

9. After examining all the columns, set $\Theta_s^{(3)} = \widehat{\Theta}_s^{(3)}$

10. Increment s. If $s \leq t_f/2$, then return to (ii). Otherwise, continue

11. Find the optimal sequence from $\min_{i,j} \Omega'^{(3)}_{t_f/2,i} \Theta^{(3)}_{t_f/2,j}$

∎

Numerical Results for Discrete-Time Optimal Control

Using the parameter values presented in Table 7.2, we computed the optimal switching rule to minimize the total viral load concentration at the end of the treatment for the three different scenarios.

Table 8.2 Total viral load at the end of treatment using model (7.2)

Scenario	Monotherapy	Switched on failure	SWATCH	Optimal
1	1.3692×10^4	344.25	45.35	45.35
2	3.7116×10^4	919.29	60.59	57.55
3	3.8725×10^4	1.2437×10^4	1045.27	55.56

Numerical results in Table 8.2 highlight that there is a significant difference in clinical recommendation treatments and the proactive switching. The SWATCH strategy shows very low levels in viral load compared with the usual recommendation switch on failure, because SWATCH constantly affects the two intermediate genotypes. However, the SWATCH treatment could fail when the regimens do not affect the highly resistant genotype with the same intensity. Then it is important to use the treatment that impacts the most highly resistant genotype for a longer period of time. Using an optimal control approach, we decreased the viral load to undetectable levels (≤ 50 copies/ml) for the three scenarios, and we can see that periodic switching may not be optimal as shown in Fig. 8.4B.

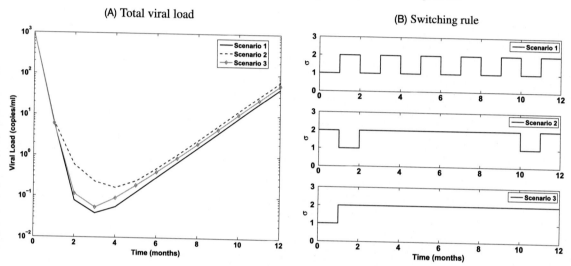

■ **FIGURE 8.4 Optimal switching treatment for model** (7.2). Three different scenarios as shown in Table 7.2 were considered.

Computational Resources

Simulation results exhibit the importance of proactive switching at the right moment to maintain viral load under detectable values. Nonetheless, computational time is a serious drawback to obtain optimal switching trajectories. One possible numerical solution is a "brute force" approach, which analyzes all possible combinations for therapies 1 and 2 with decision time $\tau = t_d$ for a time period $t_f = 336$ days, that is, $2^{\frac{t_f}{t_d}} = 2^{12}$ possible treatment combinations are evaluated. To examine long period simulations, it is necessary to find faster algorithms. In Table 8.3, different algorithms for 12 decisions are tested. It can be seen that "brute force" is extremely slow for this period of simulation, because 4096 columns are checked. Using algorithm 1, we can obtain a faster simulation by removing redundant columns. At the end of the optimization, 11 columns remain with a reduction of 99.7% columns with respect to "brute force", and computational time is reduced dramatically. Using the box constrained algorithm, this problem can be solved in less time.

Starting from initial and end points, Remark 8.9 suggests a reduction of computational time compared to the previous algorithms, that is, for every step, in both directions less columns than in a single direction algorithm are

Table 8.3 Computational resources

Method	Brute Force	Algorithm 1	Box Constraint
Time (sec)	555	4.44	3.88
Columns	4096	11	1

kept. Table 8.4 shows that the box constraint algorithm using Remark 8.9 has a lower computation time than algorithm 1. Moreover, we obtain further improvements using algorithm 3 since the process of removing columns is more effective than other algorithms due to the tighter box constraint.

Table 8.4 Computational resources using Remark 8.9

Method	Algorithm 1	Box Constraint	Algorithm 3
Time (sec)	3.3	2.3	1.74
Forward Columns	7	3	1
Backward Columns	10	6	2

These numerical examples reveal that long treatment sequences can be computed in short periods of time. For instance, using algorithm 3, the optimal trajectory for a treatment with 60 decisions is solved in 5 minutes, where nine cleaned columns are kept at the end of examination. Using "brute force", 2^{60} sequences are evaluated, something that is not possible using a standard computer. However, there is no guarantee that the proposed algorithms avoid exponential explosion in CPU time. Moreover, for a small number of decisions, the "brute force" algorithm can overperform computational times of the proposed algorithms.

Here we propose an LP problem to delete redundant columns, for implementation, using the MATLAB toolbox "linprog". For long simulation periods, the "linprog" tool alerted us to some numerical issues. These warnings can arise for different reasons, two of which are explained next.

Firstly, the unstable nature of the high resistant genotype in HIV problem causes an explosion in some states of the system, that is, if long periods are considered, then the columns of the matrix Ω start to grow exponentially, for example, 10^{28}. Then, when constraint (8.51) is checked, the "linprog" algorithm searches for very small values, that is, 10^{-24}. The default tolerance for "linprog" is 10^{-6}. Therefore this can cause numerical difficulties in the algorithm. This problem might be solved by normalizing the matrix Ω in every step. However, this solution not always works, because one of the states is always stable, whereas the other is always unstable, and then in one column different time scales can be presented. Therefore, matrix normalization might not be helpful for some problems.

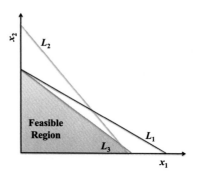

■ **FIGURE 8.5 Feasible region in the shape of a simple polygon.** L_1, L_2, and L_3 represent different constraints in the optimization problem.

Secondly, linear programming problems are solved by constructing a feasible solution at a vertex of the polytope and then walking along a path on the edges of the polytope to vertices with nondecreasing values of the objective function until an optimum is reached. However, constraints can be overly stringent and cause difficulties to the solution. If an optimization problem is considered between two variables, then we observe in Fig. 8.5 the feasible region of the problem. Note that the intersection between L_1 and L_3 is not obvious. These problems are presented in the proposed algorithms, where for some examples the intersection between lines was difficult to observe because time scales were very small (i.e., 10^{-12}). Moreover, "linprog" could have difficulties when constraints are tight, for instance, if lines L_1 and L_2 are almost parallel. An easy way to clean columns is examining element by element the columns; if all elements of one column are greater than those of the other, then it can be deleted. This may help in some examples to reduce the number of warnings and can be combined with the proposed algorithms to make them faster. A Matlab code for Algorithm 3, the fastest one, can be found in Box 8.13.

8.13 MATLAB CODE FOR ALGORITHM 3

```
clear all; close all; clc;

tic  % Start Simulation Time
n=4;  % State dimension
m=2;  % Number of matrices
cost=[1,1,1,1]';  % Cost vector
T_max=336;  % Final Time
dTsim=28;  % Simulation Time

%% Steps
N=T_max/dTsim;
t=linspace(0,T_max,N+1);

%% Set up Model parameters.
mu=1e-04;                              % Mutation rate
```

```
M=[0 1 1 0;1 0 0 1;1 0 0 1;0 1 1 0]; % Mutation pattern
death_rate=0.24;

%% Reproduction rates
rho=[0.05, 0.05; 0.27, 0.15;0.01,0.25;0.27,0.27];
R1=diag(rho(:,1));
R2=diag(rho(:,2));

%% State matrices
beta=0;
A(:,:,1)=expm((R1-death_rate*eye(size(R1))+mu*M-beta*eye(size(R1)))*dTsim);
A(:,:,2)=expm((R2-death_rate*eye(size(R2))+mu*M-beta*eye(size(R1)))*dTsim);

%% Initial Conditions for the State
Vinit=1000;
xt1=Vinit;
xt2=xt1*mu;
xt3=xt1*mu;
xt4=xt2*mu+xt3*mu;
x0=[xt1 xt2 xt3 xt4]';

%% Algorithm Initialization
Psi=x0; sequence=0;
lb = zeros(length(A),1); % Lower Bound of A
T_LP=1e-5;    % Tolerance for Linear Programming
T_kc=1+1e-7; % Tolerance to keep columns

% Keep count of how many linprog warnings
maxiterr_count=0;
linprogerr_count=0;

%% COMPUTING CLEANED COLUMNS FORWARD FOR N/2 STEPS
for i=1:(N/2)
    % Generating Psi
    [dummy,dim]=size(Psi); LastPsi=Psi(:,1:dim);
    Psi=[A(:,:,1)'*LastPsi A(:,:,2)'*LastPsi];
    % Following The Sequences
    clear c1 c2
    for r=1:dim
        c1(r,:)=[1,sequence(r,:)];   c2(r,:)=[2,sequence(r,:)];
    end
    sequence=[c1;c2];
    % Linear Programming Inequality
    Ain=-Psi'; [raws_A,columns_a]=size(Ain);
    % Computing myu
    p=0; clear myu;
    delate_columns=0; jf=0;
    %% Cycle for clear columns
    for j=1:raws_A
        % Following which column is for the new Matrix
        jp=j-delate_columns;
        % Delating column just for Optimization
        Ak=Ain;  Ak(jp,:)=[];
        [raws_Ak,columns_Ak]=size(Ak);
        b=-ones(raws_Ak,1);
        % Linear Programming Function
        options = optimset('TolFun',T_LP);
[xmin(:,j),fval,exitflag,output,lambda]=linprog(Psi(:,j)',Ak,b,[],[],lb,[],[],options);
        if exitflag == 0
            maxiterr_count=maxiterr_count+1;
        end
        if exitflag < 0
            linprogerr_count=linprogerr_count+1;
        end
        % Miu Computation
        myu(j)=Psi(:,j)'*xmin(:,j);
        % Cleaning Columns for the next step
        if ((myu(j))<T_kc)
            p=p+1; Psiclean(:,p)=Psi(:,j);   cleansequence(p,:)=sequence(j,:);
        end
        % Delating Columns for LP
        if ((myu(j))>=T_kc)
            Ain(jp,:)=[];
            delate_columns=delate_columns+1;
        end
    end
    % Psi and Sequence for Next Step
```

```
        Psi_before=Psi;  sequence_before=sequence;
        clear Psi sequence
        Psi=Psiclean;  sequence=cleansequence;
        clear cleansequence Psiclean LastPsi
        %% Print how many columns are at the end of the optimization
        [rf,cf]=size(Psi);
        fprintf(' The simulation is in the step %i and the columns are %i\n',i,cf)
    end
%% COMPUTING CLEANED COLUMNS BACKWARD FOR N/2 STEPS
Psik=cost; sequencek=0;
for ik=1:(N/2)
    % Generating Psi
    [dummyk,dimk]=size(Psik);
    LastPsik=Psik(:,1:dimk);
    Psik=[A(:,:,1)'*LastPsik A(:,:,2)'*LastPsik];
    % Following The Sequences
    clear ck1 ck2
    for rk=1:dimk
        ck1(rk,:)=[1,sequencek(rk,:)];  ck2(rk,:)=[2,sequencek(rk,:)];
    end
    sequencek=[ck1;ck2];
    % Linear Programming Inequality
    Aink=-Psik';  [raws_Ak,columns_ak]=size(Aink);
    % Computing myu
    pk=0;
    clear myuk;
    delete_columnsk=0;  jfk=0;
    % Cycle for clear columns
    for jk=1:raws_Ak
        % Following which column is for the new Matrix
        jpk=jk-delete_columnsk;
        % Delating column just for Optimization
        Akk=Aink;
        Akk(jpk,:)=[];
        [raws_Akk,columns_Akk]=size(Akk);
        bk=-ones(raws_Akk,1);
        % Linear Programming Function
        options = optimset('TolFun',T_LP);
[xmink(:,jk),fval,exitflag,output,lambda]=linprog(Psik(:,jk)',Akk,bk,[],[],lb,[],[],options);
        % Miu Computation
        myuk(jk)=Psik(:,jk)'*xmink(:,jk);
        % Cleaning Columns for the next step
        if ((myuk(jk))<T_kc)
            pk=pk+1;
            Psicleank(:,pk)=Psik(:,jk);
            cleansequencek(pk,:)=sequencek(jk,:);
        end
        % Delating Columns for LP
        if ((myuk(jk))>=T_kc)
            Aink(jpk,:)=[];
            delete_columnsk=delete_columnsk+1;
        end
    end
    %% Psi and Sequence for Next Step
    Psi_beforek=Psik; sequence_beforek=sequencek;
    clear Psik sequencek
    Psik=Psicleank;    sequencek=cleansequencek;
    clear cleansequencek Psicleank LastPsik
    %% Print how many columns are at the end of the optimization
    [rfk,cfk]=size(Psik);
    fprintf(' The simulation is in the step %i and the columns are %i\n',ik,cfk)
end

%% COUPLED FORWARD AND BACKWARDS SEQUENCES
[r_opt,c_opt]= size(Psi);  [r_optk,c_optk]= size(Psik);
gp=0;
for i=1:c_opt
    for j=1:c_optk
        gp=gp+1;
        cost_final(gp)=Psi(:,i)'*Psik(:,j);
        eye_sequence(gp,:)=[i,j];
    end
end

%% Looking the Optimal Coupled Sequence
[val_opt,seq_opt]=min(cost_final);
%% Optimal Sequence
opt_seq=eye_sequence(seq_opt,:);
optimal_control=[fliplr(sequence(opt_seq(1),1:(N/2))),  sequencek(opt_seq(2),1:(N/2))];
clc;
%% Printing Simulation Time in Seconds
display('--------------------------------------------------------------- ');
display('Simulation Time'); toc
display('--------------------------------------------------------------- ');
%% Printing the total number of columns
```

```
fprintf('   The total of clean columns in the Forward is %i \n ',c_opt);
fprintf('   The total of clean columns in the Backward is %i\n ',c_optk);
fprintf('   The total of clean columns for %i steps is %i\n ',N,(c_opt*c_optk));
display('------------------------------------------------------------');
%% Printing the total Viral Load in the last Step
fprintf('   The total Viral Load in the step %i is %f \n ',N, val_opt);
display('------------------------------------------------------------');
%% Print some things to do with errors
fprintf('No. of lin prog errors is %i \n',linprogerr_count);
fprintf('No. of lin prog max iteration warnings is %i \n',maxiterr_count);
%% Plotting Optimal Control
figure(1); clf;
stairs(t(:,1:N),optimal_control);
ylabel('\sigma'); xlabel('Time (days)'); axis([0,T_max,0,3]);
title('Optimal Control');
```

Results in Matlab Command Window:

Simulation Time
Elapsed time is 3.311878 seconds.

The total of clean columns in the Forward is 7
The total of clean columns in the Backward is 10
The total of clean columns for 12 steps is 70

The total Viral Load in the step 12 is 42.561663

No. of lin prog errors is 0
No. of lin prog max iteration warnings is 0

8.7 CONCLUDING REMARKS

In this chapter, we considered the optimal control on the fixed-horizon problem. Using a generalized solution of the HJE for continuous and discrete time, we obtain optimal solutions. We pointed out that the solution results in a two-point boundary value problem, which cannot be solved using regular integration techniques.

Under certain symmetry assumptions in the replication rate and the mutation graph, results reveal that the optimal control on this class of positive switched systems is given by the Filippov trajectory along the plane $x_2 = x_3$. We introduced necessary conditions for optimality in a more general permutation problem, and the solution also remains in a sliding surface. Such behavior suggests that in the absence of other practical constraints, rapid switching between therapies may be desirable.

The numerical solution of the optimal control problem can result in an exponential growth in computational demands. Different algorithms are proposed to try to avoid the "curse of dimensionality". These strategies were based on an LP problem to remove redundant columns. Using forward and backward approaches, these algorithms relieve computational time issues.

From the clinical point of view, we can conclude that it is very important to change therapy in the right moment to impact the appearance of high resistant genotypes. Switching on failure is a conservative treatment, which could be improved using a proactive switching as was previously proposed in the SWATCH treatment.

Robustness tests showed that periodic treatment has good performance in most cases examined. Notwithstanding, it could be possible that certain number of patients were not eligible for the proactive switching. This is when high resistant genotype is affected with different intensity by distinct treatments. In those cases, a specific regimen should be designed for the patient.

This chapter provides the speculative possibility that therapy alternation may sustain viral suppression to very low levels inhibiting the emergence of resistant mutant viruses. Further research will include more realistic models and cost functions that penalize the switching.

Suboptimal Therapy Scheduling

9.1 CONTROL OF SWITCHED SYSTEMS

Switched systems present interesting theoretical challenges and are important in many real-world problems [238]. Stability is a fundamental requirement for all control systems, and switched systems are not an exception. Furthermore, stability issues become very important in switched systems. For instance, switching between individually stable subsystems may cause instability. Conversely, switching between unstable subsystems may yield a stable switched system. This kind of phenomena justifies the recent interest in the area of switched systems. Stabilization of positive systems has been studied since it is problematic to fulfill the positivity constraint on the input variables [238].

The difficulty of determining optimal trajectories for switched systems has been studied by different authors [240,241,243,244]. Nonetheless, there is no general solution for the optimal control. In the previous chapter, the difficulties in both analytical or numerical solutions to the optimal control problem for a particular class of positive switched systems were addressed.

Consequently, here we consider other control options which may not exhibit optimal performance, but may achieve reasonable results close to the optimal one. To this end, we introduce a guaranteed cost algorithm associated with the optimal control problem in continuous and discrete time for a general class of switched systems in finite-time horizon [187]. In addition, we introduce the well-known model predictive control (MPC). The MPC appears to be suitable for a suboptimal application due to its robustness to disturbances and model uncertainties, and the capability of handling constraints.

9.2 CONTINUOUS-TIME GUARANTEED COST CONTROL

In this section, we extend the stabilization work proposed by [249], which provides a result on state-feedback stabilization of autonomous linear positive switched systems through piecewise linear copositive Lyapunov functions. This is accompanied by an aside result on the existence of a switching law that can guarantee an upper bound to the achievable performance over an infinite horizon. However, note that, for the mitigation of the viral escape problem discussed in the previous chapters, the system is not stabilizable. Therefore it is necessary to follow finite-time horizon strategies. For this purpose, let us take the simplex

$$\Lambda := \left\{ \lambda \in R^N : \sum_{i=1}^{N} \lambda_i = 1, \ \lambda_i \geq 0 \right\}, \tag{9.1}$$

which allows to introduce the piecewise linear Lyapunov function

$$v(x) := \min_{i=1,\ldots,N} \alpha_i' x = \min_{\lambda \in \Lambda} \left(\sum_{i=1}^{N} \lambda_i \alpha_i' x \right). \tag{9.2}$$

The Lyapunov function in (9.2) is not differentiable everywhere, and thus the need to use the upper Dini derivative D^+ [236]. In particular, let us define the set $I(x) = \{i : v(x) = \alpha_i' x\}$. Then $v(x)$ fails to be differentiable precisely for $x \in R_+^n$ such that $I(x)$ is composed of more than one element, that is, in the conjunction points of the individual Lyapunov functions $\alpha_i' x$. We denote by $\mathcal{M}$ the subclass of Metzler matrices with zero column sum, that is, all matrices $P \in R^{N \times N}$ with elements p_{ji} such that

$$p_{ji} \geq 0 \ \forall j \neq i, \quad \sum_{j=1}^{N} p_{ji} = 0 \ \forall i. \tag{9.3}$$

As a consequence, any $P \in \mathcal{M}$ has a zero eigenvalue since $c'P = 0$ for $c' = [1 \cdots 1]$. We now state the main result on the guaranteed cost control of system (8.8).

Theorem 9.1. *Finite horizon Guaranteed Cost*

Consider the linear positive switched system (8.8) and let nonnegative vectors q_i be given. Moreover, let $P \in \mathcal{M}$, and let $\{\alpha_1(t), \ldots, \alpha_N(t)\}$, $\alpha_i(t)$:

$[0, t_f] \to R_+^n$ be any positive solutions of the coupled differential inequalities

$$\dot{\alpha}_i + A_i'\alpha_i + \sum_{j=1}^{N} p_{ji}\alpha_j + q_i \preceq 0, \quad i = 1, \ldots, N, \qquad (9.4)$$

with final condition $\alpha_i(t_f) = c$ for all i. Then, the switching rule

$$\sigma(x(t)) = \underset{i=1,\ldots,N}{\text{argmin}} \, \alpha_i'(t)x(t) \qquad (9.5)$$

is such that

$$\int_0^{t_f} q_{\sigma(\tau)}' x(t)dt + x(t_f)'c \leq \min_{i=1,\ldots,N} \alpha_i'(0)x_0 \qquad (9.6)$$

Proof. Consider the Lyapunov function

$$v(x,t) = \min_{\ell=1,\ldots,N} \alpha_\ell'(t)x(t)$$

and let $i(t) = \text{argmin}_l \, \alpha_l'(t)x(t)$. Then

$$
\begin{aligned}
D^+(v(x),t) &= \min_k \left(\dot{\alpha}_k'(t) + \alpha_k'(t)A_i x \right) \leq \dot{\alpha}_i' + \alpha_i'(t)A_i x \\
&\leq -p_{ii}\alpha_i'(t)x - \sum_{j\neq i} p_{ji}\alpha_j'(t)x - q_i'x \\
&\leq -p_{ii}\alpha_i'(t)x - \sum_{j\neq i} p_{ji}\alpha_i'(t)x - q_i'x = -q_i'x.
\end{aligned}
$$

Hence, for all $\sigma(t)$,

$$D^+(v(x)) \leq -q_{\sigma(t)}'x(t),$$

which, after integration, gives

$$
\begin{aligned}
v(x(t_f)) - v(0) &= \int_0^{t_f} D^+ v(x(\tau))d\tau \\
&\leq -\int_0^{t_f} q_{\sigma(\tau)}'x(\tau)d\tau.
\end{aligned}
$$

Therefore

$$\int_0^{t_f} q_{\sigma(\tau)}'x(\tau)d\tau + c'x(t_f) \leq v(0) = \min_{i=1,\ldots,N} \alpha_i'(0)x_0.$$

This concludes the proof. □

Note that (9.4) requires a preliminary choice of the parameters p_{ij}. In particular, note that p_{ij} and α_i in Theorem 9.1 satisfy a bilinear matrix inequality. At the cost of some conservatism in the upper bound, these parameters can be reduced to a single one, say ζ, so allowing an easy search for the best ζ as far as the upper bound is concerned.

Corollary 9.1. *Let $q \in R_+^n$ and $c \in R_+^n$, and let positive vectors $\{\alpha_1, \dots, \alpha_N\}$, $\alpha_i \in R_+^n$, satisfy for some $\zeta > 0$ the modified coupled copositive Lyapunov differential inequalities*

$$\dot{\alpha}_i + A_i'\alpha_i + \zeta(\alpha_j - \alpha_i) + q_i \preceq 0 \qquad i \neq j = 1, \dots, N, \qquad (9.7)$$

with final condition $\alpha_i(t_f) = c$ for all i. Then the state-switching control given by (9.5) is such that

$$\int_0^{t_f} q_{\sigma(\tau)}'x(t)dt + c'x(t) \leq \min_{i=1,\dots,N} \alpha_i'(0)x_0 \qquad (9.8)$$

Proof. Consider any matrix p_{ij} chosen such that $p_{ii} = -\zeta$. Then

$$\zeta^{-1} \sum_{j \neq i=1}^{N} p_{ji} = 1 \qquad \forall i = 1, \dots, N. \qquad (9.9)$$

By (9.9) Eqs. (9.4) and (9.7) are equivalent, and hence the upper bound in Theorem 9.1 holds. □

The result in Corollary 9.8 is relevant to the problem of mitigation of HIV escape in the sense that the switching rule (9.5) may be easy to compute, that is, (9.7) is considered as an equation that is solved for α_i. The performance may not be optimal, but an upper bound of the cost function (9.8) can be known in advance. This is very helpful because when a treatment is computed using (9.5), in the worst-case scenario a bound in the total viral load can be obtained.

9.3 DISCRETE-TIME GUARANTEED COST CONTROL

The biological problem of mitigating HIV mutation was described in continuous time. However, in practice, measurements can only be made infrequently. For this purpose, the discrete-time switched system (8.43) is considered. Clearly, (8.44) constrains $A_{\sigma(k)}$ to jump among the N vertices of the matrix polytope $A_1, \dots, A_N$. It is assumed that the full state vector is available and the control law is a state feedback

$$\sigma(k) = u(x(k)). \qquad (9.10)$$

The control is a function $u: R^N \to \{1, ..., N\}$. Consider the simplex (9.1) and introduce the following piecewise copositive Lyapunov function:

$$v(x(k)) := \min_{i=1,...N} \alpha_i' x(k) = \min_{\lambda \in \Lambda} \sum_{i=1}^{N} \lambda_i \alpha_i' x(k). \qquad (9.11)$$

In a similar vein to continuous time, we need the class of matrices $\mathscr{M}$ consisting of all matrices $P \in R^{N \times N}$ with elements p_{ij} such that inequality (9.3) is satisfied. Consequently, we derive a sufficient condition for the existence of a switching rule that stabilizes the discrete-time system (8.43).

Theorem 9.2. *Stability Theorem in Discrete Time*

Assume that there exist a set of positive vectors $\alpha_1, ..., \alpha_N$, $\alpha_i \in R^n_+$, and $p \in \mathscr{M}$ satisfying the coupled copositive Lyapunov inequalities

$$(A_i - I)' \alpha_i + \sum_{j=1}^{N} p_{ji} \alpha_j \prec 0. \qquad (9.12)$$

The state-switching control with

$$u(x(k)) = \operatorname*{argmin}_{i=1,...,N} \alpha_i' x(k) \qquad (9.13)$$

makes the equilibrium solution $x = 0$ of system (8.43) globally asymptotically stable (in the positive orthant) with Lyapunov function $v(x(k))$ given by (9.11).

Proof. Recalling that (9.3) is valid for $P \in \mathscr{M}$ and that $\alpha_j' x(k) \geq \alpha_{\sigma(k)}' x(k)$ for all $j = i = 1, ..., N$, we have

$$
\begin{aligned}
\Delta v(k) &= v(x(k+1)) - v(x(k)) \\
&= \min_{j=1,...,N} \{\alpha_j' x(k+1)\} - \min_{j=1,...,N} \{\alpha_j' x(k)\} \\
&= \min_{j=1,...,N} \{\alpha_j' A_{\sigma(k)} x(k)\} - \min_{j=1,...,N} \{\alpha_j' x(k)\}.
\end{aligned}
$$

By the definition of $\sigma(k)$ we have $\min_{j=1,...,N} \{\alpha_j' x(k)\} = \alpha_{\sigma(k)}' x(k)$, and therefore

$$
\begin{aligned}
\Delta v(k) &\leq \alpha_{\sigma(k)}' A_{\sigma(k)} x(k) - \alpha_{\sigma(k)}' x(k) \\
&\leq \alpha_{\sigma(k)}' (A_{\sigma(k)} - I) x(k).
\end{aligned}
$$

From (9.12) with $x(k) \neq 0$ it follows that

$$\Delta v(k) \quad < -\sum_{j=1}^{N} p_{j\sigma(k)}\alpha'_j x(k)$$

$$= -p_{\sigma(k)\sigma(k)}\alpha'_{\sigma(k)}x(k) - \sum_{j\neq\sigma(k)}^{N} p_{j\sigma(k)}\alpha'_j x(k)$$

$$\leq -p_{\sigma(k)\sigma(k)}\alpha'_{\sigma(k)}x(k) - \sum_{j\neq\sigma(k)}^{N} p_{j\sigma(k)}\alpha'_{\sigma(k)}x(k)$$

$$\leq -\sum_{j=1}^{N} p_{j\sigma(k)}\alpha'_{\sigma(k)}x(k)$$

$$= 0,$$

which proves the theorem. □

In a similar vein, it is possible to ensure an upper bound on an optimal cost function. Let q_i, $i = 1, 2, \cdots, N$, be positive vectors and consider the cost function

$$J = \sum_{k=0}^{\infty} q'_{\sigma(k)}x(k). \tag{9.14}$$

Then the following result provides an upper bound of the optimal value J^o of J.

Lemma 9.1. *Upper Bound for Infinite Horizon*

Let $q_i \in R^n_+$. Assume that there exist a set of positive vectors $\{\alpha_1, ...\alpha_N\}$, $\alpha_i \in R^n_+$, and $p \in \mathcal{M}$ satisfying the coupled copositive Lyapunov inequalities

$$(A_i - I)'\alpha_i + \sum_{j=1}^{N} p_{ji}\alpha_j + q_i \prec 0 \ \forall i. \tag{9.15}$$

The state-switching control given by (9.13) makes the equilibrium solution $x = 0$ of system (8.43) globally asymptotically stable, and

$$J^o \leq \sum_{k=0}^{\infty} q'_{\sigma(k)}x(k) \leq \min_{i=1,...,N} \alpha'_i x_0. \tag{9.16}$$

Proof. If (9.15) holds, then (9.12) holds as well, so that the equilibrium point $x = 0$ for system (8.43) is globally asymptotically stable. In addition, by mimicking the proof of Theorem 9.2 we can prove that

$$\Delta v(x(k)) \quad = \quad v(x(k+1)) - v(x(k))$$
$$\leq -q'_{\sigma(k)}x(k).$$

Hence

$$\sum_{k=0}^{\infty} \Delta v(x(k)) \leq -\sum_{k=0}^{\infty} q'_{\sigma(k)}x(k)$$

$$\sum_{k=0}^{\infty} q'_{\sigma(k)}x(k) \leq v(x(0)) - v(x(\infty)),$$

and therefore

$$\sum_{k=0}^{\infty} q'_{\sigma(k)}x(k) \leq \min_{i=1,\dots,N} \alpha'_i x_0. \qquad \square$$

Remark 9.1. For fixed p_{ji}, to improve the upper bound provided by Lemma 9.1, we can minimize $\min_i \alpha'_i x_0$ over all possible solutions of the linear inequalities (9.15).

Coupled copositive Lyapunov functions can also be used to compute a lower bound of the optimal cost.

Lemma 9.2. *Lower Bound for Infinite Horizon*

Assume that there exist a set of positive vectors $\alpha_1, \dots, \alpha_N$, $\alpha_i \in R^n_+$, and $p \in \mathcal{M}$ satisfying the coupled copositive inequalities

$$(A_j - I)'\alpha_i + \sum_{m=1}^{N} p_{mi}\alpha_m + q_i \succeq 0 \ \forall i, j. \tag{9.17}$$

Then, for any state trajectory such that $x(k) \to 0$,

$$\sum_{k=0}^{\infty} q'_{\sigma(k)}x(k) \succeq \max_{i=1,\dots,N} \alpha'_i x_0. \tag{9.18}$$

Proof. Let

$$v(x(k)) = \max_i \alpha'_i x(k). \tag{9.19}$$

Then

$$\upsilon(x(k+1)) = \max_{i=1,...,N} \{\alpha_i' x(k+1)\}$$

$$= \max_{i=1,...,N} \{\alpha_i' A_{\sigma(k)} x(k)\}$$

$$\geq \left(\alpha_{\sigma(k)}' - \sum_{m=1}^{N} p_{m\sigma(k)} \alpha_m'\right) x(k) - q_{\sigma(k)}' x(k)$$

$$\geq \left(\alpha_{\sigma(k)}' - p_{\sigma(k)\sigma(k)} \alpha_{\sigma(k)}' - \sum_{m \neq \sigma(k)}^{N} p_{m\sigma(k)} \alpha_m'\right) x(k)$$

$$- q_{\sigma(k)}' x(k)$$

$$\geq \left(\alpha_{\sigma(k)}' - p_{\sigma(k)\sigma(k)} \alpha_{\sigma(k)}' - \sum_{m \neq \sigma(k)}^{N} p_{m\sigma(k)} \alpha_{\sigma(k)}'\right) x(k)$$

$$- q_{\sigma(k)}' x(k)$$

$$\geq \alpha_{\sigma(k)}' x(k) - q_{\sigma(k)}' x(k),$$

which implies

$$\upsilon(x(k+1)) - \upsilon(x(k)) \geq -q_{\sigma(k)}' x(k),$$

so that

$$\sum_{k=0}^{\infty} q_{\sigma(k)}' x(k) \geq \max_{i=1,...,N} \alpha_i' x_0.$$

$\square$

Remark 9.2. Note that inequalities (9.12) are not LMI, since the unknown parameters p_{ji} multiply the unknowns vectors α_j. If all A_i are Schur matrices, then a possible choice is $p_{ji} = 0, i, j = 1, 2, \cdots, N$, so that inequalities (9.12) are satisfied by $\alpha_i = (I - A_i)^{-1} \bar{q}_i$, where $\bar{q}_i \succ q_i$.

Remark 9.3. Lemma 9.2 may be used to guarantee an upper bound of the finite-time optimal cost

$$J_{FT} = c' x(t_f), \tag{9.20}$$

where t_f is the finite time, and $c \succeq 0$ is a weight on the final state $x(t_f)$. Assume that inequalities (9.12) are feasible. Hence, thanks to linearity of (9.12) in α, it is possible to find $\alpha_i \succeq 0$ such that (9.12) are satisfied along with the additional constraint $c \preceq \alpha_i$ for all i. Then $c' x(t_f) \leq \min_i \alpha_i' x(t_f) = \upsilon(x(t_f)) \leq \upsilon(x(0)) = \min_i \alpha_i' x(0)$.

The theorems and lemmas presented refer to a cost function over an infinite-time horizon. However, it is possible to slightly modify the relevant inequalities to account for finite-time horizon functionals. To be precise, consider system (8.43), the cost function

$$J = c'x(t_f) + \sum_{k=0}^{t_f-1} q'_{\sigma(k)}x(k), \tag{9.21}$$

and the difference equations

$$\alpha_i(k) = A'_i\alpha_i(k+1) + \sum_{j=1}^{N} p_{ji}\alpha_j(k) + q_i, \quad \alpha_i(t_f) = c, \tag{9.22}$$

for $i = 1, 2, \cdots, N$. We have the following result.

Theorem 9.3. *Finite-Horizon Guaranteed Cost Control*

Let $q_i \in R_+^n, i = 1, \ldots, N$. Let $\{\alpha_1(k), \ldots \alpha_N(k)\}$, $\alpha_i(k) \in R_+^n$, be a set of nonnegative vectors satisfying (9.22), where $p \in \mathcal{M}$. Then the state-switching control

$$\sigma(k) = \operatorname*{argmin}_{i=1,\ldots,N} \alpha'_i(k)x(k) \tag{9.23}$$

is such that

$$c'x(t_f) + \sum_{k=0}^{t_f-1} q'_{\sigma(k)}x(k) \le \min_{i=1,\ldots,N} \alpha'_i(0)x_0. \tag{9.24}$$

Proof. Let $v(x(k), k) = \min_i\{x(k)'\alpha_i(k)\}$. Then

$$
\begin{aligned}
v(x(k+1), k+1) &= \min_i\{x(k+1)'\alpha_i(k+1)\} \\
&= \min_i\{x(k)'A'_{\sigma(k)}\alpha_i(k+1)\} \\
&\le x(k)'A'_{\sigma(k)}\alpha_{\sigma(k)}(k+1) \\
&\le v(x(k), k) - x(k)'q_{\sigma(k)} - x(k)'\sum_{r=1}^{N} p_{r\sigma(k)}\alpha_r(k) \\
&\le v(x(k), k) - x(k)'q_{\sigma(k)} - x(k)'\alpha_{\sigma(k)}(k)\sum_{r=1}^{N} p_{r\sigma(k)} \\
&\le v(x(k), k) - x(k)'q_{\sigma(k)},
\end{aligned}
$$

so that

$$
\begin{aligned}
J &= c'x(t_f) + \sum_{k=0}^{t_f-1} q'_{\sigma(k)} x(k) \\
&\leq c'x(t_f) - \sum_{k=0}^{t_f-1} v(x(k+1), k+1) - v(x(k), k) \\
&\leq c'x(t_f) - v(x(t_f), t_f)) + v(x_0, 0) \\
&\leq \min_i \{x'_0 \alpha_i(0)\}. \quad \square
\end{aligned}
$$

Remark 9.4. Note that in the infinite-horizon case (9.17) may be infeasible. However, in the finite-horizon case (9.22) the equations are always feasible (e.g., taking $p_{ji} = 0$) and for any fixed t_f can be solved by the reversed-time difference equation (9.22). If A_i are Schur matrices, then in the limit and with $p_{ji} = 0$, $\lim_{t_f \to \infty} \alpha_i(0) = (I - A_i)^{-1} q_i$.

Corollary 9.2. *Let $q \in R_+^n$ and $c \in R_+^n$, and let the positive vectors $\{\alpha_1, \ldots, \alpha_N\}$, $\alpha_i \in R_+^n$, satisfy for some $\zeta > 0$ the modified coupled copositive Lyapunov difference equations*

$$
\alpha_i(k) = A'_i \alpha_i(k+1) + \zeta(\alpha_j(k) - \alpha_i(k)) + q_i, \qquad i \neq j = 1, \ldots, N,
$$
(9.25)

with final condition $\alpha_i(t_f) = c$ for all i. Then the state-switching control given by (9.23) is such that

$$
c'x(t_f) + \sum_{k=0}^{t_f-1} q'_{\sigma(k)} x(k) \leq \min_{i=1,\ldots,N} \alpha'_i(0) x_0.
$$
(9.26)

The proof of Corollary 9.2 is similar to that of Corollary 9.1. For the numerical solution of (9.25), a fixed ζ is considered.

9.4 MODEL PREDICTIVE CONTROL

MPC appeared in industry as an effective algorithm to deal with multivariable constrained control problems. Much progress has been made on feasibility of the online optimization, stability, and performance for linear systems. A thorough overview of the MPC history can be found in [250]. However, many systems are in general inherently nonlinear and, accompanied with high product specifications in the process industry, make nonlinear MPC systems a difficult problem. MPC involves solving an online finite-horizon open-loop optimal control problem subject to system dynamics and constraints involving states and controls. The essence of MPC is to optimize, over the manipulable inputs, forecasts of process behavior. The forecasting is achieved with a process model, and therefore a good model is required to represent the problem under study. However, models are never perfect, and therefore there will inevitably be some forecasting errors. Feedback can help overcome these effects [251]. A fundamental question about MPC is its robustness to model uncertainty and noise.

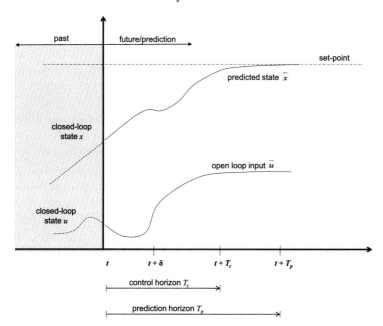

■ **FIGURE 9.1 Model Predictive Control Scheme.** T_p is the prediction horizon, whereas T_c is the control horizon [251]. The measurement is every δ sampling time units.

MPC is based on measurements obtained at time t. The controller then predicts the future dynamic behavior of the system over a prediction horizon T_p and computes an open-loop optimal control problem with control horizon T_c to generate both current and future predicted control signals. The time difference between the recalculation/measurements is fixed, that is, the measurement is every δ sampling time units. A MPC scheme is illustrated in Fig. 9.1. Due to disturbances, measurement noise and model-plant mismatch, and the true system behavior is different from the predicted one. To incorporate a feedback mechanism, the first step of the optimal control sequence is implemented. When the next measurement becomes available, at time $t + \delta$, the whole procedure – prediction and optimization – is repeated to find a new input function with the control and prediction horizons moving forward. It is clear that the shorter the horizon, the less computational time of the online optimization problem. Therefore it is desirable from a computational point of view to implement MPC using short horizons. However, when a finite prediction horizon is used, the actual closed-loop input and state trajectories will differ from the predicted open-loop trajectories even if no model plant mismatch and no disturbances are present [252].

Notice that the goal of computing a feedback such that performance objective over infinite horizon of the closed loop is not achieved. In general, a repeated minimization over a finite horizon objective in receding horizon manner by no means leads to an optimal solution for the infinite-horizon problem; in fact, the two solutions differ significantly if a short horizon is chosen. Moreover, if the predicted and actual trajectories differ, then there is no guarantee that the closed-loop system will be stable [253]. Recently, biology problems have been a productive application area for MPC [254–256]. Nevertheless, modeling in biology is a difficult task: low-order models are usually too simple to be useful, and, conversely, high-order models are too complex for simulation since they have several unknown parameters requiring identification. In this chapter, the use of MPC techniques to plan treatment applications for HIV is envisaged. This idea is not new; for example, a feedback-based treatment scheduling for HIV patients is summarized in [256]. MPC strategies have been applied to the control of HIV infection with the final goal of implementing an optimal structured treatment interruptions protocol [255]. Nevertheless, previous approaches do not accurately reflect the interaction between different genotypes and drug treatments, and consequently do not predict the possibility of the appearance of highly resistant genotypes.

Mathematical Formulation of MPC. From the biological nature of HIV infection, system (8.43) is unstable and in fact not stabilizable. This is because of the existence of a highly resistant genotype that is not affected by any treatment. Therefore, once the highly resistant mutant has "emerged", the population will explode after a period of time. The objective of MPC is to suppress the total viral load as shown in (8.23). To distinguish the real system and the system model used to predict the future for the controller, the internal variables $(\bar{x}, \bar{\sigma})$ in the controller are denoted by a bar, where $x(t) \in \mathcal{X} \subseteq R^n$ and $\sigma(\cdot) \in \mathcal{U} \subseteq R^m$. The MPC problem can be formulated as portrayed in Box 9.1.

BOX 9.1 MPC FORMULATION

Problem 9.1. Find

$$\min_{\bar{\sigma}} J(x(t), \bar{\sigma}(\bullet); T_c, T_p)$$

with

$$J(x(t), \bar{\sigma}(\bullet); T_p, T_c) := cx(t + T_p)$$

subject to:

$$\dot{\bar{x}}(\tau) = A_{\bar{\sigma}(\tau)} \bar{x}(\tau), \quad \bar{x}(\tau) = x(t),$$
$$\bar{\sigma}(\tau) \in \mathcal{U} \ \forall \tau \in [t, t + T_c],$$
$$\bar{\sigma}(\tau) = \bar{\sigma}(\tau + T_c) \ \forall \tau \in [t + T_c, t + T_p],$$
$$\bar{x}(\tau) \in \mathcal{X} \ \forall \tau \in [t, t + T_p],$$

where T_p and T_c are the prediction and the control horizon with $T_c \leq T_p$. The bar denotes internal controller variables. The distinction between the real system and the variables in the controller is necessary since the predicted values, even in the nominal undisturbed case, in generally will not be the same as the actual close-loop values, since the optimal control is recalculated at every sampling instance.

MPC Algorithm

1. Given $x(t)$ at time t, compute the open-loop optimal control $\bar{\sigma}(\cdot)$ for a receding horizon T_p.
2. Apply only the first input of the optimal command sequence to the system.
3. The remaining optimal inputs are disregarded.
4. Collect the new measurement from the system and increment t.
5. Continue with the point (1) until the final time is reached.

∎

In this algorithm, T_p has to be chosen in advance. As was previously mentioned, the shorter the horizon, the less costly the solution of the online optimization problem. The method to solve the open-loop optimal problem using (8.51) has an exponential growth. Therefore, it is desirable to use short-horizon MPC schemes for computational reasons. In general, it is not true that a repeated minimization over a finite-horizon objective in a receding horizon manner leads to an optimal solution for the infinite-horizon problem [253]. In fact, both solutions differ significantly if a short horizon is chosen.

9.5 MITIGATING HIV ESCAPE SIMULATIONS

Comparisons to the basic mutation model. To illustrate the applicability of the previous suboptimal strategies, we consider the basic mutation model presented in Chapter 7.4. The switching rule presented is computed using (9.5), where $\alpha(t)$ can be obtained from inequality (9.4), which is easily solved by considering $p_{ji} = 0$.

For Scenario 1, using a symmetric cost function weighting as $c = [1, 1, 1, 1]'$, an upper bound can be computed, and for such a control, a performance no worse than 2432.07 copies/ml is obtained. However, the final cost for a simulation period of 12 months is 1367.31 copies/ml, and this result is exactly the same as the optimal control performance.

Simulation results show that at least in some cases, guaranteed cost control captures the possible sliding-mode behavior of the optimal control law. Indeed, consider a matrix $P \in \mathcal{M}$ and its Frobenius eigenvector β, that is, such that $P\beta = 0$. It is known that β is a nonnegative vector, and it is possible to choose it such that $\sum_{i=1}^{N} \beta_i = 1$. Now it is easy to see that the solution of the differential equations (9.4) associated with the choice $\gamma \Pi \in \mathcal{M}$ is such that $\lim_{\gamma \to \infty} \alpha_i(t) = \bar{\alpha}(t)$ for all i. To characterize the limit function $\bar{\alpha}$, multiply each Eq. (9.4) by β_i and sum up all them. Since $\sum_{i=1}^{N} \beta_i p_{ji} = 0$, and $\alpha_i(t) = \bar{\alpha}(t)$, this results in

$$-\dot{\bar{\alpha}}(t) = \left(\sum_{i=1}^{N} \beta_i A_i \right) \bar{\alpha}(t) + \sum_{i=1}^{N} \beta_i q_i.$$

This equation is analogous to the equation of the costate time evolution along a sliding mode. Therefore the guaranteed cost control is capable of generating a possible sliding behavior as exhibited by the optimal trajectories satisfying, in some time interval, the equation $\dot{\bar{x}}(t) = \left(\sum_{i=1}^{N} \beta_i A_i \right) \bar{x}(t)$.

For comparison purposes, we introduced the use of MPC, which typically achieves superior performance with respect to other control strategies when manipulated and controlled variables have constraints to meet. However, for this particular application, the system is not stabilizable. The MPC objective is to delay as far as possible the escape of the system. In general, the finite set of possible control values causes problems for many control techniques. Nevertheless, in the course of MPC, having a finite set of options may be an advantage in making the optimization easier to solve.

Table 9.1 presents the performance for all proposed strategies designed to mitigate the viral escape in model (7.2). For Scenario 1, we can observe that all the strategies present an optimal behavior with the exception of the switch on failure scheme. The best switching regimen consists of switching as soon as possible for the next regimen. This recycling treatment can po-

Table 9.1 Total viral load at the end of treatment using model (7.2)

Scenario	Switch on failure	SWATCH	Optimal	Guaranteed	MPC
1	344.25	45.35	45.35	45.35	45.35
2	919.29	60.59	57.55	73.92	57.62
3	1.2437×10^4	1045.27	55.56	83.14	55.56

tentially decrease the viral load through the years and delay the appearance of resistant mutants.

For Scenario 2, the switch on failure strategy performs worse than proactive switching. For instance, the optimal strategy is very close to undetectable levels, where the detection threshold is approximately 50 copies/ml. Numerical results illustrate the good performance of MPC since the viral load achieved is very close to the optimal one. The guaranteed cost control has a good behavior for this scenario, even though its simple design is able to maintain reasonably good results when compared to the optimal with much less computational time. Scenario 3 unveils that both suboptimal strategies have good performance when compared with the optimal. Furthermore, MPC shows for this example an optimal behavior, and thus MPC can be a good strategy to deal with the mitigation of viral escape.

Comparisons for the macrophage mutation model. It was previously shown that using the macrophage mutation model (7.5), the first virological failure is presented after six years of treatment, and then the second therapy may last for five years before viral explosion is presented. Numerical results are consistent with clinical observation. In addition the SWATCH treatment outperformed the switch on failure treatment. This motivates the study of a more structured method to design a treatment regimen aimed at minimizing the viral load.

Therefore the optimal and suboptimal strategies to mitigate the viral escape in HIV are explored. Using 10 months as prediction horizon for MPC, the closed-loop response of T_i, M_i, V and the switched drug therapy are computed over a period of 20 years. The MPC algorithm achieves the goal of containing the viral load for a long period. For this example, MPC has the same performance as SWATCH. This coincidence suggests that oscillating between one treatment and the other will improve the viable treatment duration. Table 9.2 shows the total viral load using different control strategies. The suboptimal strategies have a similar performance to the optimal strategy. For instance, the guaranteed cost control slightly outperforms MPC and SWATCH. Using the guaranteed cost control, we can calculate in advance an upper bound of the performance achieved by the controller. Using optimal and suboptimal strategies, we can delay the appearance of a virologic failure by approximately three years compared to the switch on failure strategy.

Table 9.2 Results for 7-year treatment period

Strategy	Viral Load
Virologic Failure	780
SWATCH	25.8
Optimal Control	9.9
Guaranteed Cost	13.38
MPC	25.53

Comparisons for the latently infected CD4+ T cell model. In the latently infected CD4+ T cell model, three different therapies are simulated. Using the virologic failure treatment, there are three changes in therapy in a period of 4 years. The first therapy keeps the viral concentration below 1000 copies/ml for one year; however, resistant genotypes appear. Therefore the second treatment is introduced, where the viral population is again decreased for a while, but because of the existence of a high concentration of infected cells, virologic failure occurs in a shorter time period. It is then necessary to introduce the third treatment. An important fact is that latently infected cells remain almost constant, which fits with clinical and theoretical studies showing that these cells play an important role for the late stage in HIV infection. Using 10 months as prediction horizon for MPC, the closed-loop response of T_i, L, V and the switched drug therapy over a period of 14 year are computed. The MPC algorithm accomplishes the main goal of containing the viral load as long as possible.

In fact, for this example, virologic failure would be present after 12 years, which means that it could extend the duration of virologic control by 8 years compared to the clinical assessment. In addition the total population of infected CD4+ T cells is decreased for a period of 6 years. These results are consistent with the preliminary clinical trial SWATCH [229], which concluded that proactive alternation of antiretroviral regimens might extend the long-term effectiveness of treatment options without adversely affecting patients' adherence or quality of life.

Table 9.3 shows that the oscillating drug regimen provides very close results to both MPC and optimal control. Based on simulation results and clinical trials, oscillating drugs may minimize the emergence of drug-resistant strains better than frequent monitoring for viral rebound. Moreover, the guaranteed cost strategy requires very low computational resources and exhibits good performance with respect to a switch on virologic failure treatment protocol.

Table 9.3 Results for 5-year
treatment period

Strategy	Viral Load
Virologic Failure	5.14e28
SWATCH	6.22e-9
Optimal Control	2.20e-9
Guaranteed Cost	2.53e-4
MPC	2.31e-9

9.6 **CONCLUDING REMARKS**

In this chapter, alternatives to the optimal control problem for positive switched linear systems were presented. Specifically, the guaranteed cost control and MPC were implemented for mitigating HIV escape. Using a piecewise copositive Lyapunov function, sufficient conditions for stability in positive switched linear systems for continuous and discrete time were derived.

Both continuous and discrete time can provide a switching rule that guarantees a bound of the achieved performance. This means that the worst achievement of the controller can be known in advance. At the cost of some conservatism in the upper bound, the parameters in the controller can be reduced to a single one, allowing an easy computation line search for a single parameter. Due to the biological application, a switching treatment for a specific period of time was developed. Therefore, the guaranteed cost control to a finite horizon is extended. Numerical results revealed that guaranteed cost control achieved good results with minimal computational resources. Moreover, this strategy could exhibit an optimal performance for some symmetric cases and results very close to the optimal for some asymmetric cases.

The MPC application was explored to mitigate the viral escape. Simulation results exhibit the effectiveness of this method. MPC gave similar performance to the optimal control with the advantage that MPC requires less computational resources than the optimal strategy and can also handle constraints.

PK/PD-based Impulsive Control

CONTENTS

10.1 **INTRODUCTION**

Several pharmaceutical companies have taken a strategic initiative to promote the use of modeling approaches within drug projects; see Box 10.1. The value of a model-based approach for improving efficiency and decision making during the preclinical stage of drug development has been largely advocated [5]. Besides mathematical modeling, control theoretical approaches have been employed to design and schedule treatments for infectious diseases such as influenza [257] and HIV [247,256,258].

BOX 10.1 PK/PD MODELING

Drug administration considers mainly two phenomena, the **pharmacokinetics** (PK) and **pharmacodynamics** (PD). PK is the study of the temporal distribution, metabolism, and excretion of a drug in different organs of the host: what the body does to a drug [259]. The PD describes the effect of a drug on the organism: what a drug does to the body [259].

In previous chapters, optimal and suboptimal approaches were developed to schedule treatment in HIV infection neglecting the PK/PD phases of a drug. Merging PK/PD dynamics modeling with mechanistic infectious diseases models has risen as a promising tool to quantify infectious diseases, to identify potential therapeutic targets [30], and to enhance vaccines [57].

Among different control techniques, impulsive control seems to be very suitable for scheduling therapies in different medical applications [257,258, 260]. **Impulsive systems** describe processes with at least one impulsively changeable state variable. A complete reference for impulsive systems can be found in [261]. The design of optimal control policies in impulsive systems is a complex task. To relax the solution for the Hamilton–Jacobi–Bellman equation, a meaningful cost functional can be proposed a posteriori in the inverse optimal problem. Thus a stabilizing feedback control law is proposed, and then the cost functional is derived [262].

10.2 INVERSE OPTIMAL IMPULSIVE CONTROL

This subsection briefly discusses the optimal control methodology and its properties. Consider the general discrete-time nonlinear dynamical system of the form

$$x(k+1) \quad = \quad f(x(k)) + gu(k) \tag{10.1}$$

with $x_0 = x(0)$ as the equilibrium point, where $x(k) \in \mathbb{R}^n$ is the state of the system in time $k \in \mathbb{N}$, $u(k) \in \mathbb{R}^m$ is the input, and $f : \mathbb{R}^n \to \mathbb{R}^n$ is a smooth mapping with $f(0) = 0$. We consider a general class of nonlinear systems where g is a constant input gain vector. The optimal control case in which g is a state-dependent mapping in a nonlinear system model can be found in [261].

The discrete-time system (10.1) represents a system whose plant $f(x(k))$ is controlled by $u(k)$ on each kth time step. This means that the control action is present in all the dynamic time steps. However, there are certain cases in which it is not possible to keep a control input over all time steps of the system. In fact, for that kind of cases, the control input is restricted to perform at some of the kth instants. This fact helps to establish a description of the inverse optimal impulsive control for discrete-time systems.

Considering the impulsive control theory [261], system (10.1) can be expressed as

$$x(k+1) \quad = \quad f(x(k)), \ x_0 = x(0), \ k \neq \tau_k, \tag{10.2}$$

$$\Delta x(\tau_k) \quad = \quad gu(\tau_k), \ k = \tau_k, \tag{10.3}$$

$$\Delta x(\tau_k) \quad = \quad x(k+1) - x(\tau_k), \tag{10.4}$$

$$x(k+1) \quad = \quad x_0^+, \tag{10.5}$$

where $k \in \mathbb{N}$, and $\tau_k \in S$ is the impulsive instant of control. For analysis purposes, suppose that the discrete-time impulsive system (10.2)–(10.3) at the instant τ_k is represented by

$$x(k+1) \quad = \quad f(x(\tau_k)) + gu(\tau_k), \ k = \tau_k, \tag{10.6}$$

whose form is employed later in the analysis. Based on the optimal control theory [261], it is desired to determine an impulsive control law $u(\tau_k) = \bar{u}(x_i(\tau_k))$ that minimizes the cost functional of the form

$$W(x_i(\tau_k)) \;=\; \sum_{n=\tau_k}^{\infty}(l(x_i(n)) + u(n)^T Ru(n)), \qquad (10.7)$$

where $W : \mathbb{R}^n \to \mathbb{R}^+$ is a performance measure [234], $l : \mathbb{R}^n \to \mathbb{R}^+$ is a positive semidefinite function weighting the performance of the state vector $x_i(\tau_k)$, and $R : \mathbb{R}^n \to P^m$ is a real symmetric and positive definite matrix weighting the control force [234].

Remark 10.1. A function $l(z)$ is a positive semidefinite (or nonnegative definite) function if $l(z) \geqslant 0$ for all vectors z. There may be vectors z for which $l(z) = 0$, and $l(z) > 0$ for all remaining z.

Remark 10.2. A real symmetric matrix R is positive definite if $z^T Rz > 0$ for all $z \neq 0$.

Eq. (10.7) can be divided as follows:

$$\begin{aligned} W(x_i(\tau_k)) \;=\;& l(x_i(\tau_k)) + u(\tau_k)^T Ru(\tau_k) \\ &+ \sum_{n=\tau_{k+1}}^{\infty}(l(x_i(n)) + u(n)^T Ru(n)), \\ =\;& l(x_i(\tau_k)) + u(\tau_k)^T Ru(\tau_k) + W(x_i(\tau_{k+1})). \quad (10.8) \end{aligned}$$

Using the optimality principle of Bellman [234], for the infinite-horizon optimization case, the value function $W^*(x_i(\tau_k))$ becomes invariant in time. This satisfies the discrete-time Bellman equation, which can be solved backward in time [234]. This equation can be written as follows:

$$W^*(x_i(\tau_k)) = \min_{u(\tau_k)}\{l(x_i(\tau_k)) + u(\tau_k)^T Ru(\tau_k) + W^*(x_i(\tau_{k+1}))\}. \quad (10.9)$$

In addition, the Hamiltonian function $\mathcal{H}(x_i(\tau_k), u(\tau_k))$ is established to set the conditions that the optimal control law is desired to satisfy as follows [261]:

$$\begin{aligned} \mathcal{H}(x_i(\tau_k), u(\tau_k)) \;=\;& l(x_i(\tau_k)) + u(\tau_k)^T Ru(\tau_k) + W^*(x_i(\tau_{k+1})) \\ &- W^*(x_i(\tau_k)), \qquad (10.10) \end{aligned}$$

which is employed to obtain the control law $u(\tau_k)$ by calculating

$$\min_{u(\tau_k)} \mathcal{H}(x_i(\tau_k), u(\tau_k)). \qquad (10.11)$$

The minimization of (10.11) is achieved with a feedback control law denoted as $u(\tau_k) = \bar{u}(x_i(\tau_k))$; as a consequence,

$$\min_{u(\tau_k)} \mathcal{H}(x_i(\tau_k), u(\tau_k)) \quad = \quad \mathcal{H}(x_i(\tau_k), \bar{u}(x_i(\tau_k))). \qquad (10.12)$$

The feedback optimal control law in (10.12) needs to satisfy the following necessary condition [234]:

$$\mathcal{H}(x_i(\tau_k), \bar{u}(x_i(\tau_k))) \quad = \quad 0.$$

Finally, $\bar{u}(x_i(\tau_k))$ is obtained by calculating the gradient of (10.10) with respect to $u(\tau_k)$ [262], that is,

$$
\begin{aligned}
\frac{\partial W^*(x_i(\tau_k))}{\partial u(\tau_k)} \quad &= \quad \frac{\partial (l(x_i(\tau_k)) + u(\tau_k)^T R u(\tau_k))}{\partial u(\tau_k)} \\
&\quad + \frac{\partial x_i(\tau_{k+1})}{\partial u(\tau_k)} \frac{\partial W^*(x_i(\tau_{k+1}))}{\partial x_i(\tau_{k+1})} = 0 \\
&= \quad 2Ru(\tau_k) + (0 + g^T) \frac{\partial W^*(x_i(\tau_{k+1}))}{\partial x_i(\tau_{k+1})} \\
&= \quad 2Ru(\tau_k) + g^T \frac{\partial W^*(x_i(\tau_{k+1}))}{\partial x_i(\tau_{k+1})},
\end{aligned}
$$

from which it follows that

$$
\begin{aligned}
2Ru(\tau_k) \quad &= \quad -g^T \frac{\partial W^*(x_i(\tau_{k+1}))}{\partial x_i(\tau_{k+1})}, \\
u(\tau_k) \quad &= \quad -\frac{1}{2}R^{-1}g^T \frac{\partial W^*(x_i(\tau_{k+1}))}{\partial x_i(\tau_{k+1})}. \qquad (10.13)
\end{aligned}
$$

Eq. (10.13) represents the feedback optimal control law $\bar{u}(x_i(\tau_k))$ acting over each τ_kth time with $\bar{u}(0) = 0$. Note that $u(x_i(\tau_k)) = \bar{u}(x_i(\tau_k))$ is used only to indicate the optimal control. The boundary condition $W(0) = 0$ in (10.7) and (10.8) is satisfied for $W(x_i(\tau_k))$ [262]. A feedback impulsive control law is adapted from [261,262]. The solution of the inverse optimal control can be obtained with a quadratic control Lyapunov function (CLF).

Corollary 10.1. *Consider the discrete-time impulsive controlled system (10.6). The optimal control law*

$$u(\tau_k)^* = -\frac{1}{2}R^{-1}g^T \frac{\partial W^*(x_i(\tau_{k+1}))}{\partial x_i(\tau_{k+1})} \qquad (10.14)$$

is inverse optimal if

(i) It achieves exponential stability (globally) of the equilibrium point $x_i(\tau_k) = 0$ *for system (10.6);*

(ii) It minimizes the cost functional (10.7), for which $l(x_i(\tau_k)) := -\overline{W}$ *and*

$$\overline{W} := u(\tau_k)^{*T} R u(\tau_k)^* + W(x_i(\tau_{k+1})) - W(x_i(\tau_k)) \leqslant 0. \quad (10.15)$$

See further details in [262].

From Corollary 10.1 we can propose a function based on $W(x_i(\tau_k))$ such that Definition 10.1 is guaranteed. In this way, solving the HJB equation is not necessary; instead, we consider a function in quadratic form such as

$$W(x_i(\tau_k)) = \frac{1}{2}x_i(\tau_k)^T P x_i(\tau_k) \ P = P^T > 0 \quad (10.16)$$

which is applied to the optimal control law (10.14), achieved by selecting a correct matrix P. At this point the inverse optimal impulsive control law can be established combining (10.14) with (10.16), which optimizes a cost functional of the form (10.7). The control law (10.14), considering Eq. (10.6), is modified as

$$
\begin{aligned}
u(\tau_k)^* &= -\frac{1}{2}R^{-1}g^T\frac{\partial W^*(x_i(\tau_{k+1}))}{\partial x_i(\tau_{k+1})} \\
&= -\frac{1}{2}R^{-1}g^T\frac{\partial}{\partial x_i(\tau_{k+1})}(\frac{1}{2}x_i(\tau_{k+1})^T P x_i(\tau_{k+1})) \\
&= -\frac{1}{2}R^{-1}g^T P x_i(\tau_{k+1}) \\
&= -\frac{1}{2}R^{-1}g^T P f(x_i(\tau_k)) - \frac{1}{2}R^{-1}g^T P g u(\tau_k)^*,
\end{aligned}
$$

from which we have

$$u(\tau_k)^* + \frac{1}{2}R^{-1}g^T P g u(\tau_k)^* = -\frac{1}{2}R^{-1}g^T P f(x_i(\tau_k)),$$

$$(I + \frac{1}{2}R^{-1}g^T P g)u(\tau_k)^* = -\frac{1}{2}R^{-1}g^T P f(x_i(\tau_k)),$$

and, by multiplying by R both sides of the last relation,

$$(R + \frac{1}{2}g^T P g)u(\tau_k)^* = -\frac{1}{2}g^T P f(x_i(\tau_k)).$$

Finally, for $\alpha(x_i(\tau_k)) := u(\tau_k)^*$,

$$\alpha(x_i(\tau_k)) = -\frac{1}{2}(R + P_\alpha)^{-1}P_\beta(x_i(\tau_k)), \quad (10.17)$$

$$P_\alpha = \frac{1}{2}g^T P g,$$

$$P_\beta(x_i(\tau_k)) = g^T P f(x_i(\tau_k)).$$

Eq. (10.17) is the inverse optimal impulsive control law. Note that the existence of the inverse matrix in (10.17) is guaranteed since P_α is a positive definite symmetric matrix. In addition, the matrix R can be selected as the identity matrix of proper size [261,262]. The control law (10.17) establishes the input to the dynamical system (10.6) at the instant τ_k.

10.3 TAILORING INFLUENZA TREATMENT

For prevention and treatment of acute infections, antiviral drugs are an important adjunct to vaccines. For instance, the most common antiviral drugs to treat influenza infections are neuraminidase inhibitors, for example, Zanamivir, Peramivir, and Oseltamivir. The viral neuraminidase is an enzyme found on the virus surface enabling influenza virions to be released from the host infected cell. The neuraminidase inhibitors block this activity, thus interfering with viral spread and infectivity in the lungs [263]. In humans, the drug reduces clinical symptoms by 0.7–1.5 days when treatment is started 2 days after laboratory-confirmed influenza, representing great potential if used appropriately to prevent the development of resistance [264].

In this section, we employ the model proposed by Boianelli et al. [8], which was presented in Subsection 4.3. In addition, the model includes the drug dynamics (D) covering the PK/PD drug phases that provide a framework for linking interactions between drugs. The complete model is formed with the following differential equations:

$$\dot{E}(t) = S_E + r E(t)\left(\frac{V(t)}{V(t)+k_e}\right) - c_e E(t), \tag{10.18}$$

$$\dot{V}(t) = p(1-\eta)V(t)\left(1 - \frac{V(t)}{k_v}\right) - c_v V(t)E(t), \tag{10.19}$$

$$\dot{D}(t) = -\delta_D D(t), \quad \tau_k \le t < \tau_{k+1}. \tag{10.20}$$

The PK phase, the temporal distribution of drug concentration in the host, is described by Eq. (10.20). The PD phase, the effect of a drug on the organism, can be expressed by

$$\eta = \frac{D(t)}{D(t) + EC_{50}}, \tag{10.21}$$

where EC_{50} is the drug concentration level at which the 50% of drug efficacy is provided. The initial state values are $E(0) = 10^6$ cells, $V(0) =$

25 PFU/mL, and $D(0) = 0$ mg. Influenza infection model parameters are described in Subsection 4.3. Drug intake is given with $\tau_k \leq t < \tau_{k+1}$ indicating the time intervals between drug intakes. This indicates that the first drug intake of the treatment is given by $D(\tau_1) = D(t_0)$. PK/PD parameter values are EC_{50} (42.30 mg) and δ_D (3.26 days^{-1}) [257].

For therapy design and comparison, it is important to define suitable performance indicators. Thus we consider the **virological efficacy index** Ψ defined as

$$\Psi = 100 \left(1 - \frac{AUC_c}{AUC} \right) \%, \tag{10.22}$$

where AUC_c represents the area under the dynamics of virus with treatment, and AUC is the area under the curve of virus without treatment [265]. Furthermore, the *total drug amount* used is recorded for different scheduling approaches, which is the sum of all administrated drugs (mg).

Simulations. The control action in (10.3) performs with the impulsive control law $u(x_i(\tau_k)) = \alpha(x_i(\tau_k))$ defined in (10.17). The control law considers the drug state $x_D(\tau_k)$ as the only variable to be modified, based on this fact, $g = [0, 0, 1]$. As a result, P_α and $P_\beta(x_i(\tau_k))$ in (10.17) have the following forms:

$$P_\alpha = \frac{1}{2}[0, 0, 1]^T P[0, 0, 1],$$

$$P_\beta(x_i(\tau_k)) = [0, 0, 1]^T P[x_E(\tau_k), x_V(\tau_k), x_D(\tau_k)],$$

with

$$P = \begin{bmatrix} 1.0000 & 0.0100 & -0.00003 \\ 0.0100 & 1.0000 & -0.008 \\ -0.00003 & -0.008 & 1.0000 \end{bmatrix}. \tag{10.23}$$

For comparison purposes, there was considered the FDA curative treatment consisting of 75 mg of oseltamivir twice per day [266]. It was tested two initiation days of treatment, day 2 and day 3 after infection. The treatment is active until day 8 after infection. For the impulsive controller, only the first drug intake is fixed to 75 mg, and the rest of the doses are calculated by the controller. Fig. 10.1 shows the dynamics of the within-host influenza infection with initiating medication at day 2. Note that the impulsive control doses are equivalent to those presented by the FDA. This may be attributed to the high viral load on day 2. On the other hand, after day 6 post-infection, it is notable that the impulsive control doses are reduced with respect to those by the FDA. In this case, the efficacy indices, FDA ($\Psi = 93\%$) and control ($\Psi = 92.37\%$), are slightly different. However, the total amounts of drug are different, FDA (900 mg) and impulsive control (677 mg).

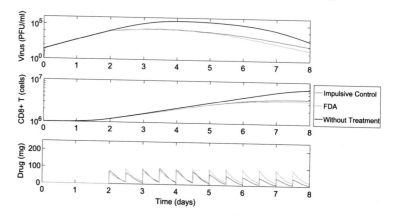

■ FIGURE 10.1 Impulsive control and FDA doses. The virus, the immune CD8+ T cells, and drug dynamics are shown. The treatment initiates at day 2. The system without therapy is also plotted.

The impulsive control performance can be evaluated under a population approach. This framework allows testing the robustness of the impulsive control. Monte Carlo (MC) simulations are used for treatment initiating at days 1, 2, and 3. The established set of parameters is represented by

$$MC_N = \{E(0), V(0), EC_{50}, \delta_D, p, k_v, c_v, S_E, k_e, c_e\}.$$

The values in MC_N were varied with a uniform random distribution within the interval of $\pm 20\%$. 1000 simulations per day are considered. A 100 mg dose bound for the impulsive controller is provided to avoid an overdose input. The Monte Carlo simulation results are depicted in Fig. 10.2, showing the drug efficacy index Ψ for each initiation day of treatment and the total amount of drug. The mean of 1000 simulations and a 95% confidence interval error bar depict the performance of each technique. For the three different initiation days, the means of treatment efficacy are slightly different in both FDA and the impulsive control. On the other hand, the total drug amount means have notable differences. Consistently with the scenario of initiating treatment on day 1, the impulsive control drug mean is less than the FDA mean, with a difference of 294 mg in this case. Note that when initiating treatment at day 3, the confidence interval of drug efficacy is larger than in the other scenarios. This effect is present in both the FDA and impulsive control treatments. However, the impulsive control efficacy variability (the confidence interval range) is smaller than the variability of the FDA medication.

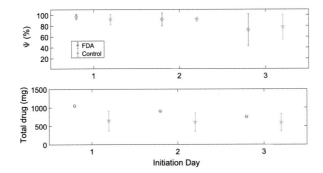

■ **FIGURE 10.2 Impulsive control and FDA efficacy indicator Ψ and total drug amount.** For each initiation day, the mean of the efficacy index and total drug are shown. The error bars show the 95% confidence interval.

10.4 **CONCLUDING REMARKS**

This chapter presents an integration of the influenza model to interact with the oseltamivir PK/PD dynamics. For the three initiation days, the mean of impulsive control doses is smaller than the fixed doses of FDA therapy. Importantly, the efficacy index of FDA treatment is reached by the impulsive control scheme in all the tested cases. Moreover, the variability of the efficacy is smaller with the impulsive control compared to the FDA treatment in the three cases. The impulsive control showed a robust behavior for the ±20% of parameter variation range and the 1000 simulations per day for the three treatment initiation scenarios tested. When treatment strategies are initiating 4 days post-infection, both strategies provide poor viral efficacy. Further techniques that allow us to schedule therapies at different time intervals will help us to advance steps toward clinical practices. There are several aspects to improve in impulsive therapies. First, observers design to viral infections as those presented in [267] can be pivotal to tackle the limited measurements in clinical practices. Moreover, bioinspired optimization schemes can serve to optimal fine-tune impulsive control parameters.

Bibliography

[1] WHO | disease outbreak news. http://www.who.int/csr/don/en/, 2016.

[2] Microbiology by numbers. Nature Reviews Microbiology 2011;9(9):628.

[3] World Health Organization. WHO | antimicrobial resistance. Tech. rep., 2016. http://www.who.int/mediacentre/factsheets/fs194/en/.

[4] WHO fact sheet; global HIV situation. http://www.who.int/gho/hiv/en/, 2015.

[5] Visser SA, Aurell M, Jones RD, Schuck VJ, Egnell AC, Peters SA, Brynne L, Yates JW, Jansson-Löfmark R, Tan B, Cooke M, Barry ST, Hughes A, Bredberg U. Model-based drug discovery: implementation and impact. Drug Discovery Today 2013;18(15–16):764–75.

[6] Tomlin CJ, Axelrod JD. Biology by numbers: mathematical modelling in developmental biology. Nature Reviews Genetics 2007;8(5):331–40.

[7] Beauchemin C, Handel A. A review of mathematical models of influenza A infections within a host or cell culture: lessons learned and challenges ahead. BMC Public Health 2011;11(Suppl 1):S7.

[8] Boianelli A, Nguyen VK, Ebensen T, Schulze K, Wilk E, Sharma N, Stegemann-Koniszewski S, Bruder D, Toapanta FR, Guzmán C, Meyer-Hermann M, Hernandez-Vargas EA. Modeling influenza virus infection: a roadmap for influenza research. Viruses 2015;7(10):5274–304.

[9] Perelson AS, Guedj J. Modelling hepatitis C therapy: predicting effects of treatment. Nature Reviews Gastroenterology and Hepatology 2015;12(8):437–45.

[10] Nguyen VK, Klawonn F, Mikolajczyk R, Hernandez-Vargas EA. Analysis of practical identifiability of a viral infection model. PLOS ONE 2016:e0167568.

[11] Murphy K, Weaver C. Janeway's immunobiology. 8th ed. Garland Science. ISBN 978-0-8153-4505-3, 2016. 888 pp.

[12] Ferrero-Miliani L, Nielsen OH, Andersen PS, Girardin SE. Chronic inflammation: importance of NOD2 and NALP3 in interleukin-1beta generation. Clinical and Experimental Immunology 2007;147(2):227–35.

[13] Bevan MJ. Helping the CD8+ T-cell response. Nature Reviews Immunology 2004;4(8):595–602.

[14] Murphy KM, Reiner SL. The lineage decisions of helper T cells. Nature Reviews Immunology 2002;2(12):933–44.

[15] Blum KS, Pabst R. Lymphocyte numbers and subsets in the human blood. Do they mirror the situation in all organs?. Immunology Letters 2007;108(1):45–51.

[16] Carter J, Saunders V. Virology: principles and applications. John Wiley & Sons. ISBN 9780470023860, 2007. p. 197.

[17] Crick FHC. Central dogma of molecular biology. Nature 1970;227(5258):561–3.

[18] Guidelines for the use of antiretroviral agents in HIV-1-infected adults and adolescents. Tech. rep., Department of Health and Human Services; 2017.

[19] Sanjuan R, Domingo-Calap P. Mechanisms of viral mutation. Cellular and Molecular Life Sciences 2016;73(23):4433–48.

[20] Jenkins GM, Rambaut A, Pybus OG, Holmes EC. Rates of molecular evolution in RNA viruses: a quantitative phylogenetic analysis. Journal of Molecular Evolution 2002;54(2):156–65.

[21] Hoenen T, Safronetz D, Groseth A, Wollenberg KR, Koita OA, Diarra B, Fall IS, Haidara FC, Diallo F, Sanogo M, Sarro YS, Kone A, Togo ACG, Traore A, Kodio M, Dosseh A, Rosenke K, de Wit E, Feldmann F, Ebihara H, Munster VJ, Zoon KC, Feldmann H, Sow S. Virology. Mutation rate and genotype variation of Ebola virus from Mali case sequences. Science 2015;348(6230):117–9.

[22] Liu H, Shen L, Zhang XL, Li XL, Liang GD, Ji HF. From discovery to outbreak: the genetic evolution of the emerging Zika virus. Emerging Microbes and Infections 2016;5(10):e111.

[23] Volk SM, Chen R, Tsetsarkin KA, Adams AP, Garcia TI, Sall AA, Nasar F, Schuh AJ, Holmes EC, Higgs S, Maharaj PD, Brault AC, Weaver SC. Genome-scale phylogenetic analyses of Chikungunya virus reveal independent emergences of recent epidemics and various evolutionary rates. Journal of Virology 2010;84(13):6497–504.

[24] Robertson DL, Hahn BH, Sharp PM. Recombination in AIDS viruses. Journal of Molecular Evolution 1995;40(3):249–59.

[25] von Kleist M. Combining pharmacology and mutational dynamics to understand and combat drug resistance in HIV. Ph.D. thesis, Maynooth: National University of Ireland; 2009.

[26] Wargo AR, Kurath G. Viral fitness: definitions, measurement, and current insights. Current Opinion in Virology 2012;2(5):538–45.

[27] Duffy S, Shackelton LA, Holmes EC. Rates of evolutionary change in viruses: patterns and determinants. Nature Reviews Genetics 2008;9(4):267–76.

[28] Gershenfeld NA. The nature of mathematical modeling. Cambridge University Press. ISBN 9780521570954, 1999. 344 pp.

[29] Altrock PM, Liu LL, Michor F. The mathematics of cancer: integrating quantitative models. Nature Reviews Cancer 2015;15(12):730–45.

[30] Boianelli A, Sharma-Chawla N, Bruder D, Hernandez-Vargas EA. Oseltamivir PK/PD modeling and simulation to evaluate treatment strategies against influenza-pneumococcus coinfection. Frontiers in Cellular and Infection Microbiology 2016;6.

[31] Perelson AS, Essunger P, Cao Y, Vesanen M, Hurley A, Saksela K, Markowitz M, Ho DD. Decay characteristics of HIV-1-infected compartments during combination therapy. Nature 1997;387:188–91.

[32] Miao H, Xia X, Perelson AS, Wu H. On identifiability of nonlinear ode models and applications in viral dynamics. SIAM Review Society for Industrial and Applied Mathematics 2011;53(1):3–39.

[33] Duvigneau S, Sharma-Chawla N, Boianelli A, Stegemann-Koniszewski S, Nguyen VK, Bruder D, Hernandez-Vargas EA. Hierarchical effects of pro-inflammatory cytokines on the post-influenza susceptibility to pneumococcal coinfection. Scientific Reports 2016;6(1):37045.

[34] Yates A, Stark J, Klein N, Antia R, Callard R. Understanding the slow depletion of memory CD4+ T cells in HIV infection. PLoS Medicine 2007;4(5):0948.

[35] Lancaster P, Tismenetsky M. The theory of matrices: with applications. Academic Press. ISBN 9780124355606, 1985. 570 pp.

[36] Poznyak AS. Advanced mathematical tools for automatic control engineers. Volume 1, deterministic techniques. Elsevier. ISBN 9780080446745, 2008. 774 pp.

[37] Khalil HK. Nonlinear systems. Third ed. Upper Saddle River: Prentice Hall. ISBN 0130673897, 2002. 750 pp.

[38] May R, McLean A. Theoretical ecology. Oxford: Oxford University Press. ISBN 9780199209989, 2007.

[39] Lotka A. Contribution to the theory of the theory of reactions. Journal of Physical Chemistry 1910;14(3):271–4.

[40] Feinberg M. Lectures on chemical reactions 4, 1980.

[41] De Boer R, Perelson A. Target cell limited and immune control models of HIV infection: a comparison. Journal of Theoretical Biology 1998;190(3):201–14.

[42] Perelson AS. Modelling viral and immune system dynamics. Nature Reviews Immunology 2002;2(1):28–36.

[43] Conway JM, Perelson AS. Residual viremia in treated HIV+ individuals. PLOS Computational Biology 2016;12(1):e1004677.

[44] Baccam P, Beauchemin C, Macken CA, Hayden FG, Perelson AS. Kinetics of influenza A virus infection in humans. Journal of Virology 2006;80(15):7590–9.

[45] Nguyen VK, Binder SC, Boianelli A, Meyer-Hermann M, Hernandez-Vargas EA. Ebola virus infection modeling and identifiability problems. Frontiers in Microbiology 2015;6:1–11.

[46] Nuraini N, Tasman H, Soewono E, Sidarto KA. A with-in host Dengue infection model with immune response. Mathematical and Computer Modelling 2009;49(5–6):1148–55.

[47] Osuna CE, Lim SY, Deleage C, Griffin BD, Stein D, Schroeder LT, Omange R, Best K, Luo M, Hraber PT, Andersen-Elyard H, Ojeda EFC, Huang S, Vanlandingham DL, Higgs S, Perelson AS, Estes JD, Safronetz D, Lewis MG, Whitney JB. Zika viral dynamics and shedding in rhesus and cynomolgus macaques. Nature Medicine 2016;22(12):1448–55.

[48] Nowak MA, Bangham CR. Population dynamics of immune responses to persistent viruses. Science 1996;272(5258):74–9.

[49] Korobeinikov A. Global properties of basic virus dynamics models. Bulletin of Mathematical Biology 2004;66(4):879–83.

[50] Hernandez-Vargas EA, Wilk E, Canini L, Toapanta FR, Binder SC, Uvarovskii A, Ross TM, Guzman C, Perelson AS, Meyer-Hermann M. Effects of aging on influenza virus infection dynamics. Journal of Virology 2014;88(8):4123–31.

[51] Wu H, Zhu H, Miao H, Perelson AS. Parameter identifiability and estimation of HIV/AIDS dynamic models. Bulletin of Mathematical Biology 2008;70(3):785–99.

[52] Hill AL, Rosenbloom DIS, Fu F, Nowak MA, Siliciano RF, Faria VG, Nolte V, Schlötterer C, Teixeira L. Predicting the outcomes of treatment to eradicate the latent reservoir for HIV. Proceedings of the National Academy of Sciences 2014;111(43):15597.

[53] Pinkevych M, Cromer D, Tolstrup M, Grimm AJ, Cooper DA, Lewin SR, Sogaard OS, Rasmussen TA, Kent SJ, Kelleher AD, Davenport MP. HIV reactivation from latency after treatment interruption occurs on average every 5–8 days-implications for HIV remission. PLoS Pathogens 2015;11(7):1–19.

[54] Conway JM, Perelson AS. Post-treatment control of HIV infection. Proceedings of the National Academy of Sciences 2015;6(1):201419162.

[55] Ke R, Lewin SR, Elliott JH, Perelson AS. Modeling the effects of vorinostat in vivo reveals both transient and delayed HIV transcriptional activation and minimal killing of latently infected cells. PLOS Pathogens 2015;11(10):e1005237.

[56] Nguyen VK, Hernandez-Vargas EA. Parameter estimation in mathematical models of viral infections using R. In: Influenza virus. Methods in molecular biology, vol. 1836. 2018. p. 531–49.

[57] Nguyen VK, Hernandez-Vargas EA. Windows of opportunity for Ebola virus infection treatment and vaccination. Scientific Reports 2017;7(1):8975.

[58] Storn R, Price K. Differential evolution—a simple and efficient heuristic for global optimization over continuous spaces. Journal of Global Optimization 1997:341–59.

[59] Mullen K, Ardia D, Gil D, Windover D, Cline J. DEoptim: an R package for global optimization by differential evolution. Journal of Statistical Software 2011;40(6):1–26.

[60] Soetaert K, Cash J, Mazzia F. Differential equations. In: Solving differential equations in R. Berlin, Heidelberg: Springer; 2012. p. 1–13.

[61] Toapanta FR, Ross TM. Impaired immune responses in the lungs of aged mice following influenza infection. Respiratory Research 2009;10(1):112.

[62] Hastie T, Tibshirani R, Friedman J. The elements of statistical learning: data mining, inference and prediction. Springer. ISBN 9780387848570, 2001. 745 pp.

[63] Raue A, Kreutz C, Maiwald T, Bachmann J, Schilling M, Klingmuller U, Timmer J. Structural and practical identifiability analysis of partially observed dynamical models by exploiting the profile likelihood. Bioinformatics 2009;25(15):1923–9.

[64] Xia X, Moog C. Identifiability of nonlinear systems with application to HIV/AIDS models. IEEE Transactions on Automatic Control 2003;48(2):330–6.

[65] Ljung L, Glad T. On global identifiability for arbitrary model parametrizations. Automatica 1994;30(2):265–76.

[66] Tunali E, Tarn Tzyh-Jong. New results for identifiability of nonlinear systems. IEEE Transactions on Automatic Control 1987;32(2):146–54.

[67] Mammen E. When does bootstrap work?. Springer-Verlag. ISBN 9781461229506, 1992. 196 pp.

[68] Thiebaut R, Guedj J, Jacqmin-Gadda H, Chene G, Trimoulet P, Neau D, Commenges D. Estimation of dynamical model parameters taking into account undetectable marker values. BMC Medical Research Methodology 2006;6(38):1–9.

[69] Heldt FS, Frensing T, Pflugmacher A, Gröpler R, Peschel B, Reichl U. Multiscale modeling of influenza A virus infection supports the development of direct-acting antivirals. PLoS Computational Biology 2013;9(11):e1003372.

[70] Dobrovolny HM, Reddy MB, Kamal MA, Rayner CR, Beauchemin CAA. Assessing mathematical models of influenza infections using features of the immune response. PLoS ONE 2013;8(2):e57088.

[71] WHO | FluNet. http://www.who.int/influenza/gisrs_laboratory/flunet/en/, 2017.

[72] Kilbourne ED. Influenza pandemics of the 20th century. http://wwwnc.cdc.gov/eid/article/12/1/05-1254_article.htm, 2006.

[73] Hensley SE. Challenges of selecting seasonal influenza vaccine strains for humans with diverse pre-exposure histories. Current Opinion in Virology 2014;8:85–9.

[74] Compans RW, Oldstone MBA. Influenza pathogenesis and control—volume I. Springer. ISBN 9783319111551, 2014.

[75] de Wit E, Rasmussen AL, Feldmann F, Bushmaker T, Martellaro C, Haddock E, Okumura A, Proll SC, Chang J, Gardner D, Katze MG, Munster VJ, Feldmann H. Influenza virus A/Anhui/1/2013 (H7N9) replicates efficiently in the upper and lower respiratory tracts of cynomolgus macaques. mBio 2014;5(4):e01331-14.

[76] Van Reeth K. Cytokines in the pathogenesis of influenza. Veterinary Microbiology 2000;74(1–2):109–16.

[77] Valkenburg SA, Rutigliano JA, Ellebedy AH, Doherty PC, Thomas PG, Kedzierska K. Immunity to seasonal and pandemic influenza A viruses. Microbes and Infection 2011;13(5):489–501.

[78] Lindsley WG, Noti JD, Blachere FM, Thewlis RE, Martin SB, Othumpangat S, Noorbakhsh B, Goldsmith WT, Vishnu A, Palmer JE, Clark KE, Beezhold DH. Viable influenza a virus in airborne particles from human coughs. Journal of Occupational and Environmental Hygiene 2015;12(2):107–13.

[79] White DO, Cheyne IM. Early events in the eclipse phase of influenza and parainfluenza virus infection. Virology 1966;29(1):49–59.

[80] Pinilla LT, Holder BP, Abed Y, Boivin G, Beauchemin CAA. The H275Y neuraminidase mutation of the pandemic A H1N1 influenza virus lengthens the eclipse phase and reduces viral output of infected cells, potentially compromising fitness in ferrets. Journal of Virology 2012;86(19):10651–60.

[81] Tamura Si, Kurata T. Defense mechanisms against influenza virus infection in the respiratory tract mucosa. Japanese Journal of Infectious Diseases 2004;57(6):236–47.

[82] Miao H, Hollenbaugh JA, Zand MS, Holden-Wiltse J, Mosmann TR, Perelson AS, Wu H, Topham DJ. Quantifying the early immune response and adaptive immune response kinetics in mice infected with influenza A virus. Journal of Virology 2010;84(13):6687–98.

[83] Steel J, Lowen A. Influenza A virus reassortment. In: Current topics in microbiology and immunology influenza pathogenesis and control – volume I, vol. 385. Cham: Springer; 2014. p. 377–93.

[84] Beauchemin CAA, McSharry JJ, Drusano GL, Nguyen JT, Went GT, Ribeiro RM, Perelson AS. Modeling amantadine treatment of influenza A virus in vitro. Journal of Theoretical Biology 2008;254(2):439–51.

[85] Bocharov GA, Romanyukha AA. Mathematical model of antiviral immune response III. Influenza A virus infection. Journal of Theoretical Biology 1994;167(4):323–60.

[86] Canini L, Carrat F. Population modeling of influenza A/H1N1 virus kinetics and symptom dynamics. Journal of Virology 2011;85(6):2764–70.

[87] Cao P, Yan AWC, Heffernan J, Petrie S, Moss RG, Carolan LA, Guarnaccia TA, Kelso A, Barr IG, McVernon J, Laurie KL, McCaw JM. Innate immunity and the inter-exposure interval determine the dynamics of secondary influenza virus infection and explain observed viral hierarchies. PLOS Computational Biology 2015:1–28.

[88] Chen SC, You SH, Liu CY, Chio CP, Liao CM. Using experimental human influenza infections to validate a viral dynamic model and the implications for prediction. Epidemiology and Infection 2012;140(9):1557–68.

[89] Dobrovolny HM, Baron MJ, Gieschke R, Davies BE, Jumbe NL, Beauchemin CA. Exploring cell tropism as a possible contributor to influenza infection severity. PLoS ONE 2010;5(11):e13811.

[90] Hancioglu B, Swigon D, Clermont G. A dynamical model of human immune response to influenza A virus infection. Journal of Theoretical Biology 2007;246(1):70–86.

[91] Handel A, Longini IM, Antia R. Neuraminidase inhibitor resistance in influenza: assessing the danger of its generation and spread. PLoS Computational Biology 2007;3(12):2456–64.

[92] Handel A, Antia R. A simple mathematical model helps to explain the immunodominance of CD8 T cells in influenza A virus infections. Journal of Virology 2008;82(16):7768–72.

[93] Holder BP, Simon P, Liao LE, Abed Y, Bouhy X, Beauchemin CAA, Boivin G. Assessing the in vitro fitness of an oseltamivir-resistant seasonal A/H1N1 influenza strain using a mathematical model. PloS ONE 2011;6(3):e14767.

[94] Holder BP, Beauchemin CA. Exploring the effect of biological delays in kinetic models of influenza within a host or cell culture. BMC Public Health 2011;11(SUPPL. 1):S10.

[95] Lee HY, Topham DJ, Park SY, Hollenbaugh J, Treanor J, Mosmann TR, Jin X, Ward BM, Miao H, Holden-Wiltse J, et al. Simulation and prediction of the adaptive immune response to influenza A virus infection. Journal of Virology 2009;83(14):7151–65.

[96] Mohler L, Flockerzi D, Sann H, Reichl U. Mathematical model of influenza A virus production in large-scale microcarrier culture. Biotechnology and Bioengineering 2005;90(1):46–58.

[97] Paradis EG, Pinilla LT, Holder BP, Abed Y, Boivin G, Beauchemin CA. Impact of the H275Y and I223V mutations in the neuraminidase of the 2009 pandemic influenza virus in vitro and evaluating experimental reproducibility. PLoS ONE 2015;10(5):e0126115.

[98] Pawelek KA, Huynh GT, Quinlivan M, Cullinane A, Rong L, Perelson AS. Modeling within-host dynamics of influenza virus infection including immune responses. PLoS Computational Biology 2012;8(6):e1002588.

[99] Petrie SM, Guarnaccia T, Laurie KL, Hurt AC, McVernon J, McCaw JM. Reducing uncertainty in within-host parameter estimates of influenza infection by measuring both infectious and total viral load. PLoS ONE 2013;8(5):e64098.

[100] Price I, Mochan-Keef ED, Swigon D, Ermentrout GB, Lukens S, Toapanta FR, Ross TM, Clermont G. The inflammatory response to influenza A virus (H1N1): an experimental and mathematical study. Journal of Theoretical Biology 2015;374:83–93.

[101] Reperant LA, Kuiken T, Grenfell BT, Osterhaus A. The immune response and within-host emergence of pandemic influenza virus. Lancet 2014;384(9959):2077–81.

[102] Saenz RA, Quinlivan M, Elton D, MacRae S, Blunden AS, Mumford JA, Daly JM, Digard P, Cullinane A, Grenfell BT, McCauley JW, Wood JLN, Gog JR. Dynamics of influenza virus infection and pathology. Journal of Virology 2010;84(8):3974–83.

[103] Schulze-Horsel J, Schulze M, Agalaridis G, Genzel Y, Reichl U. Infection dynamics and virus-induced apoptosis in cell culture-based influenza vaccine production—flow cytometry and mathematical modeling. Vaccine 2009;27:2712–22.

[104] Smith AM, Adler FR, Ribeiro RM, Gutenkunst RN, McAuley JL, McCullers JA, Perelson AS. Kinetics of coinfection with influenza A virus and Streptococcus pneumoniae. PLoS pathogens 2013;9(3):e1003238.

[105] Smith AM, Smith AP. A critical, nonlinear threshold dictates bacterial invasion and initial kinetics during influenza. Scientific Reports 2016;6:38703.

[106] Tridane A, Kuang Y. Modeling the interaction of cytotoxic T lymphocytes and influenza virus infected epithelial cells. Mathematical Biosciences and Engineering: MBE 2010;7(1):171–85.

[107] Larson EW, Dominik JW, Rowberg AH, Higbee GA. Influenza virus population dynamics in the respiratory tract of experimentally infected mice. Infection and Immunity 1976;13(2):438–47.

[108] Dobrovolny HM, Gieschke R, Davies BE, Jumbe NL, Beauchemin CAA. Neuraminidase inhibitors for treatment of human and avian strain influenza: a comparative modeling study. Journal of Theoretical Biology 2011;269(1):234–44.

[109] Doherty PC, Turner SJ, Webby RG, Thomas PG. Influenza and the challenge for immunology. Nature Immunology 2006;7(5):449–55.

[110] Handel A, Longini IM, Antia R. Towards a quantitative understanding of the within-host dynamics of influenza A infections. Journal of the Royal Society, Interface/the Royal Society 2010;7(42):35–47.

[111] Smith AM, Perelson AS. Influenza A virus infection kinetics: quantitative data and models. Wiley Interdisciplinary Reviews Systems Biology and Medicine 2011;3(4):429–45.

[112] Taubenberger JK, Morens DM. The pathology of influenza virus infections. Annual Review of Pathology 2008;3:499–522.

[113] McCullers JA. The co-pathogenesis of influenza viruses with bacteria in the lung. Nature Reviews Microbiology 2014;12(4):252–62.

[114] Ortqvist A, Hedlund J, Kalin M. Streptococcus pneumoniae: epidemiology, risk factors, and clinical features. Seminars in Respiratory and Critical Care Medicine 2005;1(212):563–74.

[115] Mina MJ, McCullers JA, Klugman KP. Live attenuated influenza vaccine enhances colonization of Streptococcus pneumoniae and Staphylococcus aureus in mice. mBio 2014;5(1):e01040-13.

[116] Scheller J, Chalaris A, Schmidt-Arras D, Rose-John S. The pro- and anti-inflammatory properties of the cytokine interleukin-6. Biochimica Et Biophysica Acta 2011;1813(5):878–88.

[117] McCullers JA, Rehg JE. Lethal synergism between influenza virus and Streptococcus pneumoniae: characterization of a mouse model and the role of platelet-activating factor receptor. The Journal of Infectious Diseases 2002;186(3):341–50.

[118] Siegel SJ, Roche AM, Weiser JN. Influenza promotes pneumococcal growth during coinfection by providing host sialylated substrates as a nutrient source. Cell Host and Microbe 2014;16(1):55–67.

[119] Sun K, Metzger DW. Inhibition of pulmonary antibacterial defense by interferon-gamma during recovery from influenza infection. Nature Medicine 2008;14(5):558–64.

[120] Durando P, Iudici R, Alicino C, Alberti M, De Florentis D, Ansaldi F, Icardi G. Adjuvants and alternative routes of administration towards the development of the ideal influenza vaccine. Human Vaccines 2011;7(SUPPL.):29–40.

[121] Hsieh YC, Wu TZ, Liu DP, Shao PL, Chang LY, Lu CY, Lee CY, Huang FY, Huang LM. Influenza pandemics: past, present and future. Journal of the Formosan Medical Association 2006;105(1):1–6.

[122] Clegg CH, Rininger JA, Baldwin SL. Clinical vaccine development for H5N1 influenza. Expert Review of Vaccines 2013;12(7):767–77.

[123] Ebola haemorrhagic fever in Sudan, 1976. Report of a WHO/International Study Team, Tech. Rep. 2; 1978. http://www.ncbi.nlm.nih.gov/pubmed/307455, 1978;56(2):247–70.

[124] CDC. Ebola virus disease report. Tech. rep., 2014. www.cdc.gov/mmwr/preview/mmwrhtml/mm6339a5.htm.

[125] Bwaka MA, Bonnet M, Calain P, Colebunders R, De Roo A, Guimard Y, Katwiki KR, Kibadi K, Kipasa MA, Kuvula KJ, Mapanda BB, Massamba M, Mupapa KD, Muyembe-Tamfum J, Ndaberey E, Peters CJ, Rollin PE, Van den Enden E. Ebola hemorrhagic fever in Kikwit, Democratic Republic of the Congo: clinical observations in 103 patients. The Journal of Infectious Diseases 1999;179(s1):S1–7.

[126] Feldmann H, Jones S, Klenk HD, Schnittler HJ. Ebola virus: from discovery to vaccine. Nature Reviews Immunology 2003;3(8):677–85.

[127] Swanepoel R, Leman PA, Burt FJ, Zachariades NA, Braack LE, Ksiazek TG, Rollin PE, Zaki SR, Peters CJ. Experimental inoculation of plants and animals with Ebola virus. Emerging Infectious Diseases 1996;2(4):321–5.

[128] Leroy EM, Epelboin A, Mondonge V, Pourrut X, Gonzalez JP, Muyembe-Tamfum JJ, Formenty P. Human Ebola outbreak resulting from direct exposure to fruit bats in Luebo, Democratic Republic of Congo, 2007. Vector-Borne and Zoonotic Diseases 2009;9(6):723–8.

[129] CDC. 2014–2016 Ebola outbreak in West Africa | Ebola hemorrhagic fever | CDC. https://www.cdc.gov/vhf/ebola/outbreaks/2014-west-africa/index.html, 2016.

[130] Mahanty S, Gupta M, Paragas J, Bray M, Ahmed R, Rollin PE. Protection from lethal infection is determined by innate immune responses in a mouse model of Ebola virus infection. Virology 2003;312(2):415–24.

[131] Halfmann P, Kim JH, Ebihara H, Noda T, Neumann G, Feldmann H, Kawaoka Y. Generation of biologically contained Ebola viruses. Proceedings of the National Academy of Sciences of the United States of America 2008;105(4):1129–33.

[132] Royston P, Altman DG. Regression using fractional polynomials of continuous covariates: parsimonious parametric modelling. Applied Statistics 1994;43(3):429.

[133] Rawlins EL, Hogan BLM. Ciliated epithelial cell lifespan in the mouse trachea and lung. AJP: Lung Cellular and Molecular Physiology 2008;295(1):L231–4.

[134] Xue H, Miao H, Wu H. Sieve estimation of constant and time-varying coefficients in nonlinear ordinary differential equation models by considering both numerical error and measurement error. The Annals of Statistics 2010;38(4):2351–87.

[135] Brun R, Reichert P, Künsch HR. Practical identifiability analysis of large environmental simulation models. Water Resources Research 2001;37(4):1015–30.

[136] Soetaert K, Petzoldt T. Inverse modelling, sensitivity and Monte Carlo analysis in R using package FME. Journal of Statistical Software 2010;33:1–28.

[137] Nowak MA, Bonhoeffer S, Hill AM, Boehme R, Thomas HC, McDade H. Viral dynamics in hepatitis B virus infection. Proceedings of the National Academy of Sciences 1996;93(9):4398–402.

[138] Diekmann O, Heesterbeek JA, Metz JA. On the definition and the computation of the basic reproduction ratio R0 in models for infectious diseases in heterogeneous populations. Journal of Mathematical Biology 1990;28(4):365–82.

[139] Prescott JB, Marzi A, Safronetz D, Robertson SJ, Feldmann H, Best SM. Immunobiology of Ebola and Lassa virus infections. Nature Reviews Immunology 2017;17(3):195–207.

[140] Van Kerkhove MD, Bento AI, Mills HL, Ferguson NM, Donnelly CA. A review of epidemiological parameters from Ebola outbreaks to inform early public health decision-making. Scientific Data 2015;2:150019.

[141] Madelain V, Nguyen THT, Olivo A, de Lamballerie X, Guedj J, Taburet AM, Mentré F. Ebola virus infection: review of the pharmacokinetic and pharmacodynamic properties of drugs considered for testing in human efficacy trials. Clinical Pharmacokinetics 2016;33(December):1–17.

[142] Richardson T, Johnston AMD, Draper H. A systematic review of Ebola treatment trials to assess the extent to which they adhere to ethical guidelines. PLoS ONE 2017;12(1):e0168975.

[143] Cardile AP, Warren TK, Martins KA, Reisler RB, Bavari S. Will there be a cure for Ebola?. Annual Review of Pharmacology and Toxicology 2017;57(1):329–48.

[144] Pavot V. Ebola virus vaccines: where do we stand?. Clinical Immunology 2016;173:44–9.

[145] Martins KA, Jahrling PB, Bavari S, Kuhn JH. Ebola virus disease candidate vaccines under evaluation in clinical trials. Expert Review of Vaccines 2016;15(9):1101–12.

[146] Heppner DG, Kemp TL, Martin BK, Ramsey WJ, Nichols R, Dasen EJ, Link CJ, Das R, Xu ZJ, Sheldon EA, Nowak TA, Monath TP, Heppner DG, Kemp TL, Martin BK, Ramsey WJ, Nichols R, Dasen EJ, Fusco J, Crowell J, Link C, Creager J, Monath TP, Das R, Xu ZJ, Klein R, Nowak T, Gerstenberger E, Bliss R, Sheldon EA, Feldman RA, Essink BJ, Smith WB, Chu L, Seger WM, Saleh J, Borders JL, Adams M. Safety and immunogenicity of the rVSVΔG-ZEBOV-GP Ebola virus vaccine candidate in healthy adults: a phase 1b randomised, multicentre, double-blind, placebo-controlled, dose-response study. The Lancet Infectious Diseases 2017;17(8):854–66.

[147] Qiu X, Wong G, Audet J, Bello A, Fernando L, Alimonti JB, Fausther-Bovendo H, Wei H, Aviles J, Hiatt E, Johnson A, Morton J, Swope K, Bohorov O, Bohorova N, Goodman C, Kim D, Pauly MH, Velasco J, Pettitt J, Olinger GG, Whaley K, Xu B, Strong JE, Zeitlin L, Kobinger GP. Reversion of advanced Ebola virus disease in nonhuman primates with ZMapp. Nature 2014;514(7520):47–53.

[148] Marzi A, Robertson SJ, Haddock E, Feldmann F, Hanley PW, Scott DP, Strong JE, Kobinger G, Best SM, Feldmann H. EBOLA VACCINE. VSV-EBOV rapidly protects macaques against infection with the 2014/15 Ebola virus outbreak strain. Science 2015;349(6249):739–42.

[149] Henao-Restrepo AM, Camacho A, Longini IM, Watson CH, Edmunds WJ, Egger M, Carroll MW, Dean NE, Diatta I, Doumbia M, Draguez B, Duraffour S, Enwere G, Grais R, Gunther S, Gsell PS, Hossmann S, Watle SV, Kondé MK, Kéïta S, Kone S, Kuisma E, Levine MM, Mandal S, Mauget T, Norheim G, Riveros X, Soumah A, Trelle S, Vicari AS, Røttingen JA, Kieny MP. Efficacy and effectiveness of an rVSV-vectored vaccine in preventing Ebola virus disease: final results from the Guinea ring vaccination, open-label, cluster-randomised trial. The Lancet 2017;389(10068):505–18.

[150] Marzi A, Hanley PW, Haddock E, Martellaro C, Kobinger G, Feldmann H. Efficacy of vesicular stomatitis virus-Ebola virus postexposure treatment in rhesus macaques infected with Ebola virus Makona. Journal of Infectious Diseases 2016;214(suppl 3):S360–6.

[151] Jones SM, Feldmann H, Stroher U, Geisbert JB, Fernando L, Grolla A, Klenk HD, Sullivan NJ, Volchkov VE, Fritz EA, Daddario KM, Hensley LE, Jahrling PB, Geisbert TW. Live attenuated recombinant vaccine protects nonhuman primates against Ebola and Marburg viruses. Nature Medicine 2005;11(7):786–90.

[152] De Silva NS, Klein U. Dynamics of B cells in germinal centres. Nature Reviews Immunology 2015;15(3):137–48.

[153] Dahlke C, Kasonta R, Lunemann S, Krahling V, Zinser ME, Biedenkopf N, Fehling SK, Ly ML, Rechtien A, Stubbe HC, Olearo F, Borregaard S, Jambrecina A, Stahl F, Strecker T, Eickmann M, Lütgehetmann M, Spohn M, Schmiedel S, Lohse AW, Becker S, Addo MM, Addo MM, Becker S, Krähling V, Agnandji ST, Krishna S, Kremsner PG, Brosnahan JS, Bejon P, Njuguna P, Siegrist CA, Huttner A, Kieny MP, Modjarrad K, Moorthy V, Fast P, Savarese B, Lapujade O. Dose-dependent T-cell dynamics and cytokine cascade following rVSV-ZEBOV immunization. EBioMedicine 2017;19:107–18.

[154] Hattori T, Lai D, Dementieva IS, Montano SP, Kurosawa K, Zheng Y, Akin LR, Swist-Rosowska KM, Grzybowski AT, Koide A, Krajewski K, Strahl BD, Kelleher NL, Ruthenburg AJ, Koide S. Antigen clasping by two antigen-binding sites of an exceptionally specific antibody for histone methylation. Proceedings of the National Academy of Sciences 2016;113(8):2092–7.

[155] Li J, Duan HJ, Chen HY, Ji YJ, Zhang X, Rong YH, Xu Z, Sun LJ, Zhang JY, Liu LM, Jin B, Zhang J, Du N, Su HB, Teng GJ, Yuan Y, Qin EQ, Jia HJ, Wang S, Guo TS, Wang Y, Mu JS, Yan T, Li ZW, Dong Z, Nie WM, Jiang TJ, Li C, Gao XD, Ji D, Zhuang YJ, Li L, Wang LF, Li WG, Duan XZ, Lu YY, Sun ZQ, Kanu AB, Koroma SM, Zhao M, Ji JS, Wang FS. Age and Ebola viral load correlate with mortality and survival time in 288 Ebola virus disease patients. International Journal of Infectious Diseases 2016;42:34–9.

[156] Barre-Sinoussi F, Chermann J, Rey F, Nugeyre M, Chamaret S, Gruest J, Dauguet C, Axler-Blin C, Vezinet-Brun F, Rouzioux C, Rozenbaum W, Montagnier L. Isolation and transmission of human retrovirus (human t-cell leukemia virus). Science 1983;219(4586):856–9.

[157] Jeffrey AM. A control theoretic approach to HIV/AIDS drug dosage design and timing the initiation of therapy; 2006, (July).

[158] Naif HM. Pathogenesis of HIV infection. Infectious Disease Reports 2013;5(SUPPL. 1):26–30.

[159] Jamieson BD, Douek DC, Killian S, Hultin LE, Scripture-Adams DD, Giorgi JV, Marelli D, Koup RA, Zack JA. Generation of functional thymocytes in the human adult. Immunity 1999;10(5):569–75.

[160] Ye P, Kirschner DE, Kourtis AP. The thymus during HIV disease: role in pathogenesis and in immune recovery. Current HIV Research 2004;2(2):177–83.

[161] Clark DR, Ampel NM, Hallett CA, Yedavalli VR, Ahmad N, DeLuca D. Peripheral blood from human immunodeficiency virus type 1-infected patients displays diminished T cell generation capacity. The Journal of Infectious Diseases 1997;176(3):649–54.

[162] Schnittman SM, Singer KH, Greenhouse JJ, Stanley SK, Whichard LP, Le PT, Haynes BF, Fauci AS. Thymic microenvironment induces HIV expression. Physiologic secretion of IL-6 by thymic epithelial cells up-regulates virus expression in chronically infected cells [published erratum appears in Journal of Immunology 1992 Feb 1;148(3):981], Journal of Immunology 1991;147(8):2553–8.

[163] Wang L, Robb CW, Cloyd MW. HIV induces homing of resting T lymphocytes to lymph nodes. Virology 1997;228(2):141–52.

[164] Wang L, Chen JJY, Gelman BB, Cloyd MW, Konig R, Cloyd MW. A novel mechanism of CD4 lymphocyte depletion involves effects of HIV on resting lymphocytes: induction of lymph node homing and apoptosis upon secondary signaling through homing receptors. Journal of Immunology 1999;162(1):268–76.

[165] Ford WL, Gowans JL. The traffic of lymphocytes. Seminars in Hematology 1969;6(1):67–83.

[166] Turville SG, Cameron PU, Handley A, Lin G, Pöhlmann S, Doms RW, Cunningham AL. Diversity of receptors binding HIV on dendritic cell subsets. Nature Immunology 2002;3(10):975–83.

[167] Kwon DS, Gregorio G, Bitton N, Hendrickson WA, Littman DR. DC-SIGN-mediated internalization of HIV is required for trans-enhancement of T cell infection. Immunity 2002;16(1):135–44.

[168] Chougnet C, Gessani S. Role of gp120 in dendritic cell dysfunction in HIV infection. Journal of Leukocyte Biology 2006;80(5):994–1000.

[169] Hazenberg MD, Hamann D, Schuitemaker H, Miedema F. T cell depletion in HIV-1 infection: how CD4+ T cells go out of stock. Nature Immunology 2000;1(4):285–9.

[170] Hazenberg MD, Otto SA, van Benthem BHB, Roos MTL, Coutinho RA, Lange JMA, Hamann D, Prins M, Miedema F. Persistent immune activation in HIV-1 infection is associated with progression to AIDS. AIDS (London, England) 2003;17(13):1881–8.

[171] Letvin NL, Walker BD. Immunopathogenesis and immunotherapy in AIDS virus infections. Nature Medicine 2003;9(7):861–6.

[172] Chun TW, Stuyver L, Mizell SB, Ehler LA, Mican JAM, Baseler M, Lloyd AL, Nowak MA, Fauci AS. Presence of an inducible HIV-1 latent reservoir during highly active antiretroviral therapy. Proceedings of the National Academy of Sciences 1997;94(24):13193–7.

[173] Finzi D, Hermankova M, Pierson T, Carruth LM, Buck C, Chaisson RE, Quinn TC, Chadwick K, Margolick J, Brookmeyer R, Gallant J, Markowitz M, Ho DD, Richman DD, Siliciano RF. Identification of a reservoir for HIV-1 in patients on highly active antiretroviral therapy identification of a reservoir for HIV-1 in patients on highly active antiretroviral therapy. Science 1997;278(November):1295–300.

[174] Orenstein JM. The macrophage in HIV infection. Immunobiology 2001;204(5):598–602.

[175] Perelson AS, Neumann AU, Markowitz M, Leonard JM, Ho DD. HIV-1 dynamics in vivo: virion clearance rate, infected cell life-span, and viral generation time. Science 1996;271(5255):1582–6.

[176] Kirschner E, Webb F. Immunotherapy of HIV-1 infection. Journal of Biological Systems 1998;6(1):71–83.

[177] Bajaria SH, Webb G, Cloyd M, Kirschner D. Dynamics of naive and memory CD4+ T lymphocytes in HIV-1 disease progression. Journal of Acquired Immune Deficiency Syndromes 2002;30(1):41–58.

[178] Hogue IB, Bajaria SH, Fallert BA, Qin S, Reinhart TA, Kirschner DE. The dual role of dendritic cells in the immune response to human immunodeficiency virus type 1 infection. Journal of General Virology 2008;89(9):2228–39.

[179] Althaus CL, De Boer RJ. Dynamics of immune escape during HIV/SIV infection. PLoS Computational Biology 2008;4(7):1–9.

[180] Hadjiandreou M, Conejeros R, Vassiliadis VS. Towards a long-term model construction for the dynamic simulation of HIV infection. Mathematical Biosciences and Engineering: MBE 2007;4(3):489–504.

[181] Chang H, Astolfi A. Enhancement of the immune system in HIV dynamics by output feedback. Automatica 2009;45(7):1765–70.

[182] Hernandez-Vargas EA, Middleton RH. Modeling the three stages in HIV infection. Journal of Theoretical Biology 2013;320:33–40.

[183] Rong L, Perelson AS. Asymmetric division of activated latently infected cells may explain the decay kinetics of the HIV-1 latent reservoir and intermittent viral blips. Mathematical Biosciences 2009;217(1):77–87.

[184] Conway JM, Coombs D. A stochastic model of latently infected cell reactivation and viral blip generation in treated HIV patients. PLoS Computational Biology 2011;7(4).

[185] Zhang J, Perelson AS. Contribution of follicular dendritic cells to persistent HIV viremia. Journal of Virology 2013;87(14):7893–901.

[186] Bonhoeffer S, Nowak MA. Pre-existence and emergence of drug resistance in HIV-1 infection. Proceedings of the Royal Society B: Biological Sciences 1997;264(1382):631–7.

[187] Hernandez-Vargas EA, Colaneri P, Middleton RH, Blanchini F. Discrete-time control for switched positive systems with application to mitigating viral escape. International Journal of Robust and Nonlinear Control 2010;21(10):1093–111.

[188] Rosenbloom D, Hill A, Rabi S. Antiretroviral dynamics determines HIV evolution and predicts therapy outcome. Nature Medicine 2012;18(9):1378–85.

[189] Hernandez-Vargas EA, Colaneri P, Middleton RH. Switching strategies to mitigate HIV mutation. IEEE Transactions on Control Systems Technology 2014;(1):1–6.

[190] Moreno-Gamez S, Hill AL, Rosenbloom DIS, Petrov DA, Nowak MA, Pennings PS. Imperfect drug penetration leads to spatial monotherapy and rapid evolution of multidrug resistance. Proceedings of the National Academy of Sciences 2015;112(22):E2874–83.

[191] Fonteneau R, Stan GB, Michelet C, Zeggwagh F, Belmudes F, Ernst D, Lefebvre MA. Modelling the influence of activation-induced apoptosis of CD4+ and CD8+ T-cells on the immune system response of a HIV-infected patient. IET Systems Biology 2008;2(2):94–102.

[192] Tan WY, Wu H. Stochastic modeling of the dynamics of CD4+ T-cell infection by HIV and some Monte Carlo studies. Mathematical Biosciences 1998;147(2):173–205.

[193] Dalal N, Greenhalgh D, Mao X. A stochastic model for internal HIV dynamics. Journal of Mathematical Analysis and Applications 2008;341(2):1084–101.

[194] Zorzenon dos Santos RM, Coutinho S. Dynamics of HIV infection: a cellular automata approach. Physical Review Letters 2001;87(16):168102.

[195] Burkhead EG, Hawkins JM, Molinek DK. A dynamical study of a cellular automata model of the spread of HIV in a lymph node. Bulletin of Mathematical Biology 2009;71(1):25–74.

[196] Bergamaschi A, Pancino G. Host hindrance to HIV-1 replication in monocytes and macrophages. Retrovirology 2010;7:1–17.

[197] Brew BJ, Robertson K, Wright EJ, Churchill M, Crowe SM, Cysique LA, Deeks S, Garcia JV, Gelman B, Gray LR, Johnson T, Joseph J, Margolis DM, Mankowski JL, Spencer B. HIV eradication symposium: will the brain be left behind?. Journal of NeuroVirology 2015;21(3):322–34.

[198] Cloyd MW, Chen JJY, Wang L. How does HIV cause AIDS? The homing theory. Molecular Medicine Today 2000;6(3):108–11.

[199] Kirschner D, Webb GF, Cloyd M. Model of HIV-1 disease progression based on virus-induced lymph node homing and homing-induced apoptosis of CD4+ lymphocytes. JAIDS Journal of Acquired Immune Deficiency Syndromes 2000;24(4):352–62.

[200] Kirschner DE, Mehr R, Perelson AS. Role of the thymus in pediatric HIV-1 infection. Journal of Acquired Immune Deficiency Syndromes and Human Retrovirology 1998;18(2):95–109.

[201] Zhang L, Chung C, Hu BS, He T, Guo Y, Kim AJ, Skulsky E, Jin X, Hurley A, Ramratnam B, Markowitz M, Ho DD. Genetic characterization of rebounding HIV-1 after cessation of highly active antiretroviral therapy. Journal of Clinical Investigation 2000;106(7):839–45.

[202] Perelson AS, Nelson PW. Mathematical analysis of HIV-1 dynamics in vivo. SIAM Review 1999;41(1):3–44.

[203] Brown A, Zhang H, Lopez P, Pardo CA, Gartner S. In vitro modeling of the HIV-macrophage reservoir. Journal of Leukocyte Biology 2006;80(5):1127–35.

[204] Takahashi K, Wesselingh SL, Griffin DE, McArthur JC, Johnson RT, Glass JD. Localization of HIV-1 in human brain using polymerase chain reaction/in situ hybridization and immunocytochemistry. Annals of Neurology 1996;39(6):705–11.

[205] Greenough TC, Brettler DB, Kirchhoff F, Alexander L, Desrosiers RC, O'Brien SJ, Somasundaran M, Luzuriaga K, Sullivan JL. Long-term nonprogressive infection with human immunodeficiency virus type 1 in a hemophilia cohort. The Journal of Infectious Diseases 1999;180(6):1790–802.

[206] Fauci AS, Pantaleo G, Stanley S, Weissman D. Immunopathogenic mechanisms of HIV infection. Annals of Internal Medicine 1996;124(7):654–63.

[207] Xia X. Modelling of HIV infection: vaccine readiness, drug effectiveness and therapeutical failures. Journal of Process Control 2007;17(3):253–60.

[208] Lawn SD, Roberts BD, Griffin GE, Folks TM, Butera ST. Cellular compartments of human immunodeficiency virus type 1 replication in vivo: determination by presence of virion-associated host proteins and impact of opportunistic infection. Journal of Virology 2000;74(1):139–45.

[209] Igarashi T, Brown CR, Endo Y, Buckler-White A, Plishka R, Bischofberger N, Hirsch V, Martin MA. Macrophage are the principal reservoir and sustain high virus loads in rhesus macaques after the depletion of CD4+ T cells by a highly pathogenic simian immunodeficiency virus/HIV type 1 chimera (SHIV): implications for HIV-1 infections of humans. Proceedings of the National Academy of Sciences of the United States of America 2001;98(2):658–63.

[210] Elaiw AM. Global properties of a class of HIV models. Nonlinear Analysis: Real World Applications 2010;11(4):2253–63.

[211] Chang H, Astolfi A. Control of HIV infection dynamics. IEEE Control Systems Magazine 2008;28(2):28–39.

[212] Barão M, Lemos JM. Nonlinear control of HIV-1 infection with a singular perturbation model. IFAC Proceedings Volumes (IFAC-PapersOnline) 2006;6(PART 1):333–8.

[213] Connor EM, Sperling RS, Gelber R, Kiselev P, Scott G, O'Sullivan MJ, VanDyke R, Bey M, Shearer W, Jacobson RL, et al. Reduction of maternal-infant transmission of human immunodeficiency virus type 1 with zidovudine treatment. Pediatric AIDS clinical trials group protocol 076 study group. The New England Journal of Medicine 1994;331(18):1173–80.

[214] Cihlar T, Fordyce M. Current status and prospects of HIV treatment. Current Opinion in Virology 2016;18:50–6.

[215] Autran B. Positive effects of combined antiretroviral therapy on CD4+ T cell homeostasis and function in advanced HIV disease. Science 1997;277(5322):112–6.

[216] Department of Health and Human Services. Adults and adolescents living with HIV guidelines for the use of antiretroviral agents in adults and adolescents living with HIV. Tech. rep., 2018.

[217] Emmelkamp JM, Rockstroh JK. CCR5 antagonists: comparison of efficacy, side effects, pharmacokinetics and interactions—review of the literature. European Journal of Medical Research 2007;12(9):409–17.

[218] Ho DD. Time to hit HIV, early and hard. The New England Journal of Medicine 1995;333(7):450–1.

[219] Haggerty S, Stevenson M. Predominance of distinct viral genotypes in brain and lymph node compartments of HIV-1-infected individuals. Viral Immunology 1991;4(2):123–31.

[220] Zhang H, Dornadula G, Beumont M, Livornese L, Van Uitert B, Henning K, Pomerantz RJ. Human immunodeficiency virus type 1 in the semen of men receiving highly active antiretroviral therapy. The New England Journal of Medicine 1998;339(25):1803–9.

[221] Lisziewicz J, Rosenberg E, Lieberman J, Jessen H. Control of HIV despite the discontinuation of antiretroviral therapy. The New England Journal of Medicine 1999;340(21):1683.

[222] Harrer T, Harrer E, Kalams SA, Elbeik T, Staprans SI, Feinberg MB, Cao Y, Ho DD, Yilma T, Caliendo AM, Johnson RP, Buchbinder SP, Walker BD. Strong cytotoxic T cell and weak neutralizing antibody responses in a subset of persons with stable nonprogressing HIV type 1 infection. AIDS Research and Human Retroviruses 1996;12(7):585–92.

[223] Strategies for Management of Antiretroviral Therapy (SMART) Study Group, El-Sadr WM, Lundgren JD, Neaton JD, Gordin F, Abrams D, Arduino RC, Babiker A, Burman W, Clumeck N, Cohen CJ, Cohn D, Cooper D, Darbyshire J, Emery S, Fätkenheuer G, Gazzard B, Grund B, Hoy J, Klingman K, Losso M, Markowitz N, Neuhaus J, Phillips A, Rappoport C. CD4+ count-guided interruption of antiretroviral treatment. The New England Journal of Medicine 2006;355(22):2283–96.

[224] Ananworanich J, Gayet-Ageron A, Le Braz M, Prasithsirikul W, Chetchotisakd P, Kiertiburanakul S, Munsakul W, Raksakulkarn P, Tansuphasawasdikul S, Sirivichayakul S, Cavassini M, Karrer U, Genné D, Nüesch R, Vernazza P, Bernasconi E, Leduc D, Satchell C, Yerly S, Perrin L, Hill A, Perneger T, Phanuphak P, Furrer H, Cooper D, Ruxrungtham K, Hirschel B. CD4-guided scheduled treatment interruptions compared with continuous therapy for patients infected with HIV-1: results of the Staccato randomised trial. Lancet 2006;368(9534):459–65.

[225] Maggiolo F, Ripamonti D, Gregis G, Quinzan G, Callegaro A, Suter F. Effect of prolonged discontinuation of successful antiretroviral therapy on CD4 T cells: a controlled, prospective trial. AIDS 2004;18(3):439–46.

[226] Danel C, Moh R, Minga A, Anzian A, Ba-Gomis O, Kanga C, Nzunetu G, Gabillard D, Rouet F, Sorho S, Chaix ML, Eholié S, Menan H, Sauvageot D, Bissagnene E, Salamon R, Anglaret X. CD4-guided structured antiretroviral treatment interruption strategy in HIV-infected adults in West Africa (Trivacan ANRS 1269 trial): a randomised trial. Lancet 2006;367(9527):1981–9.

[227] Barbour JD, Wrin T, Grant RM, Martin JN, Segal MR, Petropoulos CJ, Deeks SG. Evolution of phenotypic drug susceptibility and viral replication capacity during long-term virologic failure of protease inhibitor therapy in human immunodeficiency virus-infected adults. Journal of Virology 2002;76(21):11104–12.

[228] Randomized A, Trial C, Martinez-picado J. Alternation of antiretroviral drug regimens for HIV infection. Annals of Internal Medicine 2003;139:81–9.

[229] Martinez-Cajas JL, Wainberg MA. Antiretroviral therapy: optimal sequencing of therapy to avoid resistance. Drugs 2008;68(1):43–72.

[230] D'Amato RM, D'Aquila RT, Wein LM. Management of antiretroviral therapy for HIV infection: modelling when to change therapy. Antiviral Therapy 1998;3(3):147–58.

[231] Haering M, Hordt A, Meyer-Hermann M, Hernandez-Vargas EA. Computational study to determine when to initiate and alternate therapy in HIV infection. BioMed Research International 2014;2014:472869.

[232] Sungkanuparph S, Groger RK, Overton ET, Fraser VJ, Powderly WG. Persistent low-level viraemia and virological failure in HIV-1-infected patients treated with highly active antiretroviral therapy. HIV Medicine 2006;7(7):437–41.

[233] Locatelli A. Optimal control: an introduction. Birkhäuser Verlag. ISBN 9783764364083, 2001. 294 pp.

[234] Kirk DE. Optimal control: an introduction. Prentice-Hall. ISBN 0486434842, 2004. 452 pp.

[235] Pontryagin LS. The mathematical theory of optimal processes. Gordon and Breach Science Publishers; 1986.

[236] Clarke F. Nonsmooth analysis and control theory, vol. 178. Springer. ISBN 978-0-387-98336-3, 1998.

[237] Bellman R. Dynamic programming; 2010. p. 1–56.

[238] Liberzon D. Switching in systems and control. Birkhauser. ISBN 9780817642976, 2003. 233 pp.

[239] Farina L, Rinaldi S. Positive linear systems: theory and applications, vol. 50. Wiley. ISBN 111803127X, 2011. 305 pp.

[240] Branicky MS, Borkar VS, Mitter SK. A unified framework for hybrid control: model and optimal control theory. IEEE Transactions on Automatic Control 1998;43(1):31–45.

[241] Cassandras CG, Pepyne DL, Wardi Y. Optimal control of a class of hybrid systems. IEEE Transactions on Automatic Control 2001;46(3):398–415.

[242] Xu X, Antsaklis PJ. Optimal control of switched systems based on parameterization of the switching instants. IEEE Transactions on Automatic Control 2004;49(1):2–16.

[243] Sussmann H. A maximum principle for hybrid optimal control problems. In: Proceedings of the 38th conference on decision and control, vol. 1. ISBN 0780352505, 1999. p. 425–30.

[244] Piccoli B. Necessary conditions for hybrid optimization. In: Proceedings of the 38th IEEE conference on decision and control, vol. 1. ISBN 0-7803-5250-5, 1999. p. 410–5.

[245] Bai X, Yang XS. A new proof of a theorem on optimal control of switched systems. Journal of Mathematical Analysis and Applications 2007;331(2):895–901.

[246] Murray JS, Elashoff MR, Iacono-Connors LC, Cvetkovich TA, Struble KA. The use of plasma HIV RNA as a study endpoint in efficacy trials of antiretroviral drugs. AIDS 1999;13(7):797–804.

[247] Hernandez-Vargas EA, Colaneri P, Middleton RH. Optimal therapy scheduling for a simplified HIV infection model. Automatica 2013;49(9):2874–80.

[248] Blanchini F, Miani S. Set-theoretic analysis of dynamic systems. Boston, MA: Birkhäuser Boston; 2008.

[249] Zappavigna A, Colaneri P, Geromel JC, Middleton R. Stabilization of continuous-time switched linear positive systems. In: Proceedings of the 2010 American control conference. IEEE. ISBN 978-1-4244-7427-1, 2010. p. 3275–80.

[250] Morari M, Lee JH. Model predictive control: past, present and future. Computers and Chemical Engineering 1999;23:667–82.

[251] Findeisen R, Allgower F. An introduction to nonlinear model predictive control. In: 21st Benelux meeting on systems and control; 2002. p. 121–43.

[252] Allgöwer F, Badgwell TA, Qin JS, Rawlings JB, Wright SJ. Nonlinear predictive control and moving horizon estimation – an introductory overview. In: Advances in control. London: Springer. ISBN 978-1-4471-1216-7, 1999. p. 391–449.

[253] Bitmead RR, Gevers M, Wertz V. Adaptive optimal control: the thinking man's GPC. Automatica 1993:798–800.

[254] Lynch SM, Bequette BW. Model predictive control of blood glucose in type I diabetics using subcutaneous glucose measurements. In: Proceedings of the American control conference, vol. 5. IEEE. ISBN 0780372980, 2002. p. 4039–43.

[255] Pannocchia G, Laurino M, Landi A. A model predictive control strategy toward optimal structured treatment interruptions in anti-HIV therapy. IEEE Transactions on Biomedical Engineering 2010;57(5):1040–50.

[256] Zurakowski R. Nonlinear observer output-feedback MPC treatment scheduling for HIV. Biomedical Engineering Online 2011;10(1):40.

[257] Hernandez-Mejia G, Alanis AY, Hernandez-Vargas EA. Inverse optimal impulsive control based treatment of influenza infection. In: IFAC world congress 2017, vol. 50. 2017. p. 12696–701.

[258] Rivadeneira PS, Caicedo M, Ferramosca A, Gonz AH. Impulsive Zone Model Predictive Control (iZMPC) for therapeutic treatments: application to HIV dynamics. In: 56th IEEE conference on decision and control; 2017. p. 6.

[259] Pacey S, Workman P, Sarker D. Pharmacokinetics and pharmacodynamics in drug development. In: Encyclopedia of cancer. Berlin, Heidelberg: Springer; 2011. p. 2845–8.

[260] Rivadeneira PS, Sereno JE. Blood glycemia reconstruction from discrete measurements using an impulsive observer; 2017. p. 15288–93.

[261] Wassim MH, VijaySekhar C, Sergey GN. Impulsive and hybrid dynamical systems. ISBN 0691127158, 2006. 523 pp.

[262] Sanchez EN, Ornelas-Tellez F. Discrete-time inverse optimal control for nonlinear systems. ISBN 9781466580879, 2016.

[263] Moscona A. Neuraminidase inhibitors for influenza (review). The New England Journal of Medicine 2005;353(13):1363–73.

[264] McNicholl IR, McNicholl JJ. Neuraminidase inhibitors: zanamivir and oseltamivir. Annals of Pharmacotherapy 2001;35(1):57–70.

[265] Canini L, Perelson AS. Viral kinetic modeling: state of the art. Journal of Pharmacokinetics and Pharmacodynamics 2014;41(5):431–43.

[266] WHO. WHO guidelines for pharmacological management of pandemic (H1N1) 2009 influenza and other influenza viruses. http://www.who.int/csr/resources/publications/swineflu/h1n1_use_antivirals_20090820/en/, 2009.

[267] Alanis AY, Hernandez-Gonzalez M, Hernandez-Vargas EA. Observers for biological systems. Applied Soft Computing 2014;24:1175–82.

Index

Printed in the United States
By Bookmasters